A Student Handbook for Writing in Biology

Sixth Edition

A STUDENT HANDBOOK FOR
Writing in Biology
Sixth Edition

Karin Knisely
Bucknell University

macmillan
learning

Austin • Boston • New York • Plymouth

Senior Vice President, STEM: Daryl Fox
Program Director: Sandy Lindelof
Program Manager: Lisa Lockwood
Marketing Manager: Will Moore
Executive Content Development Manager, STEM: Debbie Hardin
Executive Project Manager, Content, STEM: Katrina Mangold
Director of Content Management Enhancement: Tracey Kuehn
Senior Managing Editor: Lisa Kinne
Director of Design, Content Management: Diana Blume
Design Services Manager: Natasha Wolfe
Cover Design Manager: John Callahan
Director of Digital Production: Keri deManigold
Printing and Binding: LSC Communications

Library of Congress Control Number: 2020952594

ISBN-13: 978-1-319-30832-2
ISBN-10: 1-319-30832-5

© 2021, 2017 W. H. Freeman and Company

1 2 3 4 5 6   26 25 24 23 22 21 20

w. h. freeman
Macmillan Learning

In 1946, William Freeman founded W. H. Freeman and Company and published Linus Pauling's *General Chemistry*, which revolutionized the chemistry curriculum and established the prototype for a Freeman text. W. H. Freeman quickly became a publishing house where leading researchers can make significant contributions to mathematics and science. In 1996, W. H. Freeman joined Macmillan and we have since proudly continued the legacy of providing revolutionary, quality educational tools for teaching and learning in STEM.

Macmillan Learning
One New York Plaza
Suite 4600
New York, NY 10004-1562
www.macmillanlearning.com

*"What do we live for,
if it is not to make life less difficult for each other?"*

– Mary Ann Evans (aka George Eliot)

Contents

Preface

From its first edition, the goal of this handbook has been to provide students with a practical, readable resource for communicating their scientific knowledge according to the conventions in biology. Scientific knowledge comes from asking questions that can be answered through careful observation and measurement. Seeking information to answer those questions, designing experiments and observational studies to test hypotheses, collecting data, analyzing data, and presenting our findings to other scientists are the core components of the scientific method. For science to benefit the public, however, the information must be presented in a manner that a general audience can understand; this genre is called science communication. Because scientists are increasingly being called upon to explain their work to a general audience, several chapters in the Sixth Edition have been reworked and expanded to include science communication. Chapter 2 describes how secondary sources, such as science news articles, play an important role for gaining basic background knowledge on a research question. The role of social media in disseminating information about science and as an informal resource for scientists and the science-minded public is also covered in Chapter 2. In contrast to primary sources that have undergone peer review, information posted on social media and the internet is not checked for accuracy. Chapter 2 provides examples that show effective and ineffective ways to evaluate sources for relevance and credibility. In Chapter 3, a journal article and a science news article on the same topic are compared to illustrate the most important differences between these two genres. Documenting sources is now its own chapter (Chapter 4), and the section on citing online sources has been expanded to include blogs, podcasts, social media, among others, to reflect the growing variety of media with science content.

New to this book is a chapter on analyzing data using statistics (Chapter 6). We use an experiment on blackworms to illustrate how to explore and analyze a dataset. We show how to conduct t-tests for one sample and two sample means of numeric variables and chi-square tests for categorical variables. We describe how to propose statistical hypotheses and interpret the results to draw conclusions. Throughout the chapter, we use the programming language R in our data analysis. R is a free, open source software program that is widely used by statisticians and scientists. The chapter wraps up with a discussion of other software programs for statistical analysis and ethical concerns when analyzing data using statistics.

Objectives have been added to the beginning of each chapter to help students focus on the most important topics. Check Your Understanding questions and activities at the end of each chapter give students an opportunity to put the concepts into practice.

In the Sixth Edition, the appendices have been updated for Microsoft Office 2019 and Office 2019 for Mac. Video tutorials for both Mac and Windows, available from **macmillanlearning.com/knisely6e**, provide time-saving tips for formatting document elements in Word and applying formulas, making graphs, and saving graphs as chart templates in Excel.

The book is augmented by ancillary materials available on the Macmillan web page. A biology lab report template in Microsoft Word provides prompts that help students get used to scientific paper format and content. The Scientific Writing Checklist can be printed out to help students self-evaluate or peer review their papers. Instructors and students will find the list of proofreading marks and laboratory report comments handy for use in both the revision and feedback stages. The Evaluation Form for Oral Presentations enables listeners to provide feedback to the speaker on things that they are doing well as well as areas that need improvement. Similarly, the Evaluation Form for Poster Presentations can be used as a checklist for the presenter and an evaluation tool for visitors during the actual poster session. To illustrate principles of designing effective posters, sample posters created with poster design software are posted online; each poster is accompanied by a short evaluation of the layout and content. All of the above documents can be downloaded from **macmillanlearning.com/knisely6e**.

Acknowledgments

Many of the improvements in the Sixth Edition were made in response to feedback from reviewers of the Fifth Edition of this book. I thank James Dearworth (Lafayette College), Justin Walguarnery (University of Hawaii, Manoa), Kathryn Yurkonis (University of North Dakota), Lisa Baird (University of San Diego), and Monica Torres (Rutgers University) for their detailed and helpful suggestions.

I would like to acknowledge the colleagues and friends who reviewed chapters or provided advice in the Sixth Edition: Ken Field, Claire Kelling (Penn State University), Brian Knisely (Penn State University), Chris Martine, Lydia Naughton, Kathy Shellenberger (Midd-West High School), Rebekah Stevenson, and Tanisha Williams (Chapter 2); Dee-Ann Reeder (Chapter 3); Le Paliulis and Abby Hare-Harris (Bloomsburg University) (GWAS in Chapter 6). I thank Bucknell University students

Brianna Bjordahl, Andrei Bucaloiu, Paige Caine, Deanna Cannizzaro, Katie Edwards, Jenna Farmer, Jack Geduldig, Bitseat Getaneh, Ben Haussmann, Hannah Litwa, Lydia Naughton, Zander Perelman, Hallie Robin, Zach Sisson, Gianna Tricola, Lindsey Trusal, Jenny Waters, Abbie Winter, and their faculty mentors Morgan Benowitz-Fredericks, Mark Haussmann, Greg Pask, Mark Spiro, C. Tristan Stayton, and Mizuki Takahashi, for allowing me to display their posters on the web page.

I am especially grateful to Claire Kelling for writing the much-needed chapter on statistics. She explains how a statistician would go about analyzing biological data using R. Her contribution to this book is especially timely in view of the increasing popularity of R for data analysis in ecology, bioinformatics, and other areas of biology. Claire includes R code snippets in her analysis to allow students and instructors to repurpose the code for similar analyses. Interested readers can access the example data set and her complete R code on GitHub. On a personal note, Claire's student-centered approach, her prompt responses to my queries, and her straightforward explanations of data exploration, statistical hypothesis testing, and how to interpret the results made working with her on this project a real pleasure.

I also express my sincere appreciation to those who contributed to the previous editions of this book: reviewers Angela Currie (Niagara University), Sharon Hyak (Victoria College), Javier Izquierdo (Hofstra University), Ross E. Koning (Eastern Connecticut State University), Brenda Leicht (University of Iowa), Keenan M.L. Mack (Illinois College), Katharine Northcutt (Mercer University), Luciana Cursino Parent (Hobart and William Smith Colleges), Donald A. Stratton (University of Vermont), Amy Wiles (Mercer University), Wendy Sera (Baylor University), Elizabeth Bowdan (University of Massachusetts at Amherst), and Mary Been (Clovis Community College). I also thank John Woolsey (poster presentations), Lynne Waldman (good student lab report), Marvin H. O'Neal III (Stony Brook University), Joe Parsons (University of Victoria), Randy Wayne (Cornell University), John Byram (W.W. Norton & Co.), Kristy Sprott (The Format Group LLC), Walter H. Piper (Chapman University), Warren Abrahamson (Bucknell University), Karen Patrias (National Library of Medicine), and the staff of the Bucknell Writing Center. Debra Balducci (technology tutorials and photographing posters), Brianna Derr and Cameron Williams (production of the video tutorials), Kathleen McQuiston (article databases and scholarly search engines), and Jim Van Fleet (reference management software and scholarly research) in Library and Information Technology at Bucknell University answered my research and computer-related questions. My conversations with Katrina Knisely, Kathy Shellenberger, and Sandy Field have been invaluable in helping me clarify my ideas for multiple editions of this book.

I also want to thank my students in Introduction to Molecules and Cells (BIOL205) and Organismal Biology (BIOL206) laboratories for helping me understand the difficulties they encounter when writing lab reports. My colleagues in the Biology Department at Bucknell University continually provide me with fresh and refreshing ideas on teaching, and I am proud to be part of such a collegial group.

I am grateful to all the professionals at Sinauer Associates/Oxford University Press and Macmillan Learning who helped with the production of this book, especially Alison Hornbeck, Joan Gemme, Meg Britton Clark, Rick Neilsen, Mark Siddall, and Michele Beckta (permissions). Joan Kalkut coordinated the pre-revision reviews and was a valuable resource for me throughout the production of the Sixth Edition. Michael Jones oversaw the transfer of the digital resources to the Student Site. Andy Sinauer and Dean Scudder provided encouragement, feedback, and organizational input on the first five editions.

Finally, I would like to thank my children, Katrina, Carleton, and Brian, for their insight into what motivates today's college students and for keeping me up-to-date on the latest technological developments. Brian especially has always been willing to share his technical knowledge with me, communicating concepts with empathy and clarity. I cannot thank my late parents and my husband, Chuck, enough for their steadfast love and support. They have always encouraged me to do my best, to seek knowledge, and to share it with others. I am grateful to them for giving me the all-important time and space to pursue my goals and passions.

Karin Knisely • Lewisburg, PA
January 2021

## Chapter 1

The Scientific Method

Objectives

1.1 Understand how the scientific method is a systematic approach to answering biological questions

1.2 Summarize the steps in the scientific method

1.3 Distinguish between hypotheses and predictions

1.4 Differentiate between independent, dependent, and controlled variables

1.5 Explain the difference between experimental and statistical hypotheses

1.6 Explain some of the important considerations in designing an experiment

1.7 Describe what information should be recorded in a research notebook

1.8 Explain how a scientist might move forward after a "failed" experiment

Trying to understand natural phenomena is human nature. We are curious about why things happen the way they do, and we expect to be able to understand these events through careful observation and measurement. This process is known as the scientific method, and it is the foundation of all knowledge in the biological sciences.

1.1 An Introduction to the Scientific Method

The scientific method involves a number of steps:

- Asking questions
- Looking for sources that might help answer the questions
- Developing possible explanations (hypotheses)
- Designing an experiment to test a hypothesis
- Predicting what the outcome of an experiment will be if the hypothesis is correct
- Collecting data
- Analyzing data
- Developing possible explanations for the experimental results
- Revising original hypotheses to take into account new findings

- Designing new experiments to test new hypotheses (or other experiments to provide further support for existing hypotheses)
- Sharing findings with other scientists and the public

Most scientists do not rigidly adhere to this sequence of steps, but it provides a useful starting point for how to conduct a scientific investigation.

Ask a question

As a biology student, you are probably naturally curious about your environment. You wonder about the hows and whys of things you observe. To apply the scientific method to your questions, however, the phenomena of interest must be sufficiently well defined. The parameters that describe the phenomena must be measurable and controllable. For example, let's say that you learned that:

> Dwarf pea plants contain a lower concentration of the hormone gibberellic acid than wild-type pea plants of normal height.

You might ask the question:

> Does gibberellic acid regulate plant height?

This is a question that can be answered using the scientific method, because the parameters can be controlled and measured. On the other hand, the following question could not be answered easily with the scientific method:

> Will the addition of gibberellic acid increase a plant's sense of well-being?

In this example, "a sense of well-being" is not something that can be measured or controlled.

Look for answers to your question

There is a good chance that other people have already asked the same question. That means that there is a good chance that you may be able to find the answer to your question, if you know where to look. Secondary references such as your textbook, encyclopedias, science news articles, and information posted on the websites of university research groups, professional societies, museums, and government agencies are usually easier to comprehend than journal articles and may be good places to begin finding answers (see Section 2.1). Curiously, attempts to answer the original question often result in new questions, and unexpected findings lead to new directions in research. By reading other people's work, you may think of a more interesting question, define your question more clearly, or modify your question in some other way.

Turn your question into a hypothesis

As a result of your literature search or conversations with experts, you may now have a tentative answer to your original (or modified) question. Now it is time to develop a hypothesis. A hypothesis is a possible explanation for something you have observed. **You must have information before you can propose a hypothesis!** Without information, your hypothesis is nothing more than an uneducated guess. That is why you must look for possible answers before you can turn your question into a hypothesis.

A useful hypothesis is one that can be tested and either supported or negated. A hypothesis can never be *proven* right, but the evidence gained from your observations and/or measurements can *provide support for* the hypothesis. Thus, when scientists write papers, they never say, "The results prove that..." Instead, they write, "The results suggest that..." or "The results provide support for..."

You might transform your question "Does gibberellic acid regulate plant height?" into the following testable hypothesis:

> GOOD: The addition of gibberellic acid to dwarf
>   plants will allow them to grow to the height of
>   normal, wild-type plants.

This hypothesis provides specific expectations that can be tested. In contrast, the following hypothesis is not specific enough:

> VAGUE: The addition of gibberellic acid will affect the
>   height of dwarf plants.

Design an experiment to test your hypothesis

In an **observational study**, scientists observe individuals and measure variables of interest without trying to control the variables or influence the response. While observations provide important information about a group, it is difficult to draw conclusions about cause and effect relationships because multiple factors affect the response. That's the main reason why scientists conduct experiments. **Experiments** are studies in which the investigator imposes a specific treatment on an organism or thing (for example, a biochemical reaction) while controlling the other factors that might influence the response.

The first step in designing an experiment is to determine which variables might be influential. Of those variables, only one may be manipulated in any given experiment; the others have to remain constant. The individuals in the experiment are then divided into treatment and control groups. The treatment group is subjected to the independent variable and the control group is not; all other conditions are the same for the two groups. These experimental conditions are repeated many times to make sure the results are reliable. Statistical tests can be used to determine whether the differences between the means of the treatment and control groups are statistically significant.

The individuals in the treatment group may respond differently from those in the control group, or there may be no difference in response between the groups. These results are then related back to the hypothesis, potentially supporting or negating it.

Define the variables   Variables are commonly classified as independent or explanatory variables, dependent or response variables, and controlled variables. The *one* variable that a scientist manipulates in a given experiment is called the **independent variable** or the explanatory variable, so called because it "explains" or influences the response. It is important to manipulate *only one* variable at a time to determine whether or not a cause and effect relationship exists between that variable and an individual's response. The other variables that may affect the response must be carefully controlled so that they do not confound the relationship between the independent variable and the dependent variables.

Dependent variables are those affected by the imposed treatment; in other words, they represent an individual's response to the independent variable. Dependent variables are variables that can be measured or observed, such as size, number of seeds produced, and velocity of an enzymatic reaction.

The hypothesis proposed earlier involves testing whether there is a cause and effect relationship between gibberellic acid (GA) treatment and plant height. GA level is the variable that will be manipulated; plant height is the response that we'll measure. Because plant height is affected by many other factors such as ambient temperature, humidity, age of the plants, day length, amount of fertilizer, and watering regime, however, we must keep these controlled variables constant so that any differences in response can be attributed to the GA treatment.

Set up the treatment and control groups   The individuals in the experiment are assigned randomly to either a treatment group or a control group. Those in the treatment group will be subjected to the independent variable (GA in this case), while those in the control group will not. Depending on the hypothesis, the control group may be subdivided into positive and negative controls. Negative controls are not treated with the independent variable and are not expected to show a response. Positive controls represent a reference for treatment groups that demonstrate a response consistent with the hypothesis.

> HYPOTHESIS: Adding GA to dwarf plants will allow
> them to grow to the height of normal,
> wild-type plants.

TREATMENT GROUP: Dwarf plants + GA

CONTROL GROUPS:

    NEGATIVE: Dwarf plants + no GA (substitute an equal volume of water)

    POSITIVE: Wild-type plants + no GA

Determine the level of treatment for the independent variable   How much GA should be added to the dwarf plants in the treatment group to produce an increase in height? Too little GA may not effect a response, but too much might be toxic. To determine the appropriate level of treatment, consult the literature or carry out a preliminary experiment. The level may even be a range of concentrations that is appropriate for the biological system.

Provide enough replicates   A single result is not statistically valid. The same treatment conditions must be applied to many individuals to be confident that the results are reliable.

Make predictions about the outcome of your experiment   Predictions provide a sense of direction during both the design stage and the data analysis stage of your experiment. For each treatment and control group, predict the outcome of the experiment if your hypothesis is correct.

HYPOTHESIS: Adding GA to dwarf plants will allow them to grow to the height of normal, wild-type plants.

TREATMENT GROUP: Dwarf plants + GA

PREDICTION IF HYPO-   Dwarf plants will grow as tall as wild-type
THESIS IS SUPPORTED: plants + no GA.

NEGATIVE CONTROL: Dwarf plants + no GA

PREDICTION: Dwarf plants will be short.

POSITIVE CONTROL: Wild-type plants + no GA

PREDICTION: Wild-type plants will be tall.

If the data lend themselves to statistical testing, you would also propose statistical hypotheses, which are different from experimental hypotheses. Statistical hypotheses take the form of null and alternative hypotheses. The **null hypothesis (H_0)** says that there is no difference in response between the treatment and control groups; any observed differences can be attributed to chance. The **alternative hypothesis (H_a or H_1)** states that there is a difference in response between the treatment and control groups. For our example, the **statistical hypotheses** would be

H_0: There is no difference in the mean height of dwarf plants treated with GA and those not treated with GA.

H_1: There is a difference in the mean height of dwarf plants treated with GA and those not treated with GA.

A statistical test allows us to decide if we can reject the null hypothesis or if there is not enough evidence to reject the null hypothesis. Please see Chapter 6 for an introduction to data analysis using statistics.

Record data

Scientists record observations, procedures, and results in notebooks. The type of notebook (bound or loose leaf, with or without duplicate pages, pocket or full-size) may be prescribed by your instructor or the principal investigator of the research lab. Field notebooks, which are used by scientists who work outdoors, may also be water resistant. More important than the physical notebook, however, is the detail and accuracy of what's recorded inside. For each experiment or study, include the following information:

- Investigator's name
- The date (month, day, and year)
- The location (for field studies)
- A descriptive title
- The purpose
- The procedure (in words or as a flow chart)
- Numerical data, along with units of measurement, recorded in well-organized tables
- Drawings with dimensions and magnification, where appropriate. Structures are drawn in proportion to the whole. Parts are labeled. Observations about the appearance, color, texture, and so on are included.
- Graphs, printouts, and gel images
- Calculations
- A brief summary of the results
- Questions, possible errors, and other notes

When deciding on the level of detail, imagine that, years from now, you or someone else wants to repeat the study and confirm the results. The more information you provide, the easier it will be to understand what you did, what problems you encountered, suggestions for improving the procedure, the results you obtained, how you analyzed the data, and how you reached your conclusions.

Explore the data

The raw data in lab and field notebooks are the basis for the results published in the primary and secondary literature. Published results, however, usually represent a *summary* of the raw data by the authors. The summary usually consists of average values for each treatment and control group, along with the spread of the data. In the interest of scientific transparency, PLOS and other journal publishers are now requiring authors to make their data publicly available in repositories. This practice makes it easier for scientists to validate, replicate, and re-analyze published results.

Data exploration is a useful tool for better understanding your data. As explained in more detail in Section 6.4, data exploration can lead to the discovery of missing data or data entry errors. Graphing the data often makes it easier to spot trends and outliers.

Analyze the data

After exploring and summarizing the data, we need to determine if the results support our experimental hypothesis. For many kinds of data, the analysis involves proposing statistical hypotheses and conducting statistical tests. The test we use depends on the nature of the data (see Section 6.5). With statistical tests, we can determine if the difference between a treatment and its control is statistically significant or due to chance.

Try to explain the results

After analyzing the data, it's time to develop possible explanations for the results. You previously found information on your topic when you developed your hypothesis. Return to this material to try to explain your results. Do your results agree with those of other researchers? How did they interpret their results? If your results do not agree, try to determine why not. Did you use different methods or organisms? What were some possible sources of error?

You should realize that even some of the most elementary questions in biology have taken hundreds of scientists many years to answer. One approach to the problem may seem promising at first, but as data are collected, problems with the method or other complications may become apparent. Although the scientific method is indeed methodical, it also requires imagination and creativity. Successful scientists are not discouraged when their initial hypotheses are discredited. Instead, they are already revising their hypotheses in light of recent discoveries and planning their next experiment. You will not usually get instant gratification from applying the scientific method to a question, but you are sure to be rewarded with unexpected findings, increased patience, and a greater appreciation for the complexity of biological phenomena.

Revise original hypotheses to take new findings into account

If the data support the hypothesis, then you might design different experiments that provide additional evidence. If the data do not support the hypothesis, then it may be necessary to modify the hypothesis and carry out a new experiment. However, it should be obvious that repeating the same experiment many times just to get the desired result is unethical (see also p-hacking in Section 6.7). Ideally, scientists will thoroughly investigate a question until they are satisfied that they can explain the phenomenon of interest.

Share findings with other scientists

The final phase of the scientific method is communicating your results to other scientists, either at scientific meetings or through a publication in a journal. When you submit a paper to refereed journals, it is read critically by other scientists in your field, and your methods, results, and conclusions are scrutinized. If any errors are discovered, they are corrected before your results are communicated to the scientific community at large.

Poster sessions are an excellent way to share preliminary findings with your colleagues. The emphasis in poster presentations is on the methods and the results. The informal atmosphere promotes the exchange of ideas among scientists with common interests. See Chapter 9 on how to prepare a poster.

Oral presentations are different from both journal articles and poster sessions, because the speaker's delivery plays a critical role in the success of the communication. See Chapter 10 for tips on preparing and delivering an effective oral presentation.

Share your enthusiasm for science with the public

Communicating science to a general audience has many benefits (AAAS Communication Toolkit Fundamentals 2020). As a result of your interaction with younger students, some of them may decide to pursue a major in a STEM (science, technology, engineering, and math) field. Your ability to explain scientific concepts to community policymakers may increase their trust in science and help them make better decisions about health and the environment. See Section 3.4 for how to communicate with a general audience.

Check Your Understanding

Practice applying the scientific method.

1. Make a list of observations that pique your curiosity.
2. Write down 2-3 questions for an observation, which you want to answer.
3. Can you find answers to these questions by searching Google? Google Scholar?
4. Does your search for answers raise other questions?
5. Choose one question and turn it into a testable hypothesis.
6. How could you test your hypothesis? Identify the independent variable, the dependent variable, and the controlled variables for your experiment. If you are designing an observational study, identify the variables that you may not have control over.
7. Describe your experimental setup or the conditions of your observational study.

Go to **macmillanlearning.com/knisely6e**
and select **"Student Site"** to access
samples, template files, and tutorial videos

## Chapter 2

Finding Information About Topics in Science

Objectives

2.1 Distinguish between primary and secondary sources

2.2 Know how and when to use primary and secondary sources in your quest to find information

2.3 Evaluate the credibility and task relevance of sources based on source features and content

2.4 Understand the main differences between search engines and article databases: number of results returned, types of publications indexed, how peer review affects credibility, keyword searches versus queries

2.5 Understand how social media platforms are used to communicate scientific knowledge: pros and cons

One of the most important skills you will develop in college is information literacy. The American Library Association defines **information literacy** as "… the ability to locate, evaluate, and use effectively the needed information" (American Library Association 1989). You will search the biological literature when you write laboratory reports, research papers, research proposals, and other assignments. A vast amount of information is available on just about every topic. Finding exactly what you need is the hard part.

In your quest to find the information you need, you will encounter many types of sources that can be classified broadly as scholarly or non-scholarly (popular), primary or secondary, and digital or print, although the distinctions are not always clear-cut. Some sources are more reliable than others. Some sources are easier to read than others.

2.1 Primary and Secondary References

Secondary references are good resources for learning about a topic. Examples include textbooks, newspaper and magazine articles, Wikipedia, encyclopedias, and information posted on the websites of professional societies and government agencies. Secondary sources about science are usually written by journalists or science writers in a style that is understandable to a non-specialist audience. When pictures are present, they are primarily used to capture reader interest rather than to present experimental data.

On the other hand, **primary references** are first-hand accounts by the scientists who carried out the research. Because scientists are writing for other scientists, journal articles typically contain **jargon** (technical language used by specialists in the field) and tables and graphs, and they omit details that are known to their specialist audience. These characteristics make journal articles much harder to read than secondary sources. However, only the primary literature provides a detailed description of the methodology and the actual experimental results. This information allows scientists to repeat or build on previous studies and to compare their data with data reported by other scientists. Furthermore, primary sources contain extensive in-text citations and a full list of references at the end. Scientists use references to support their assertions and to give readers access to information that helps them develop a deeper understanding of the subject.

Primary and secondary sources play different, but equally important, roles in learning. Learning is a messy process whereby we cycle back and forth between knowing and not knowing, understanding and not understanding. We generally start out with some basic knowledge of a topic that we want to learn more about. When we encounter a primary source that provides new information, we may not be able to understand the content. Not understanding makes us realize that we need to read additional secondary sources to acquire more background knowledge. This cycle may be repeated many times in the process of learning something new.

The goal of your literature searches, especially in your upper-level biology classes, will be to find journal articles on a particular topic. As shown in Figure 2.1, however, you may need to read secondary sources at various times in your journey to acquire the basic knowledge you will need to understand the journal articles.

2.2 Overview of the Information-Seeking Process

Acquiring background information

The search for information begins with a basic understanding of your topic. To acquire background information on your topic, read your textbook and consult the resources on your library's research by subject webpages. These resources are selected specifically for the subject area by the library staff and may include links to

- Encyclopedias, dictionaries, and other reference works
- Government agencies

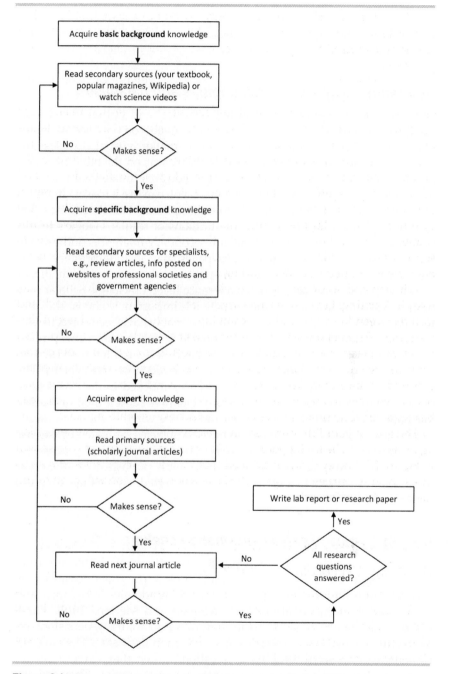

Figure 2.1   How primary and secondary sources are used at different stages of learning. To understand the content in scholarly and primary sources, we first have to acquire background information by reading popular and secondary sources.

- Professional societies
- Datasets
- Popular science

Differences between search engines and databases

Scientific information is housed in both physical and virtual libraries. To use the large, virtual public library we call the web, you only need a smartphone or a computer. Access to the much smaller, virtual libraries that house information for specialists (article databases), on the other hand, requires a subscription. Fortunately, as long as you are affiliated with a college or university that subscribes to the database, you can use it free of charge.

Google and Google Scholar are interdisciplinary **search engines** that search the entire web, often returning millions of results. Google's results pages (but not Google Scholar's) include secondary references and links to social media. Google Scholar's results pages may include primary references that are linked to scientists' homepages, professional social media, or course websites. Because a primary reference listed on Google Scholar may be an earlier version of a published article, it is important to choose the most recent version, in which any errors will presumably have been corrected.

Article databases contain a pre-screened collection of primary references and other types of documents. A database search tends to turn up fewer results, but all of the results are considered to be trustworthy. The reason is that most of the primary references indexed in databases have undergone **peer review**. In this process, scientists submit their articles to journal editors, who read the articles and decide whether the work merits publication. Usually, the editors also have the articles evaluated by experts who are not associated with the study. This quality control process ensures that the information is accurate before it is published. On the other hand, the results of a Google search are not vetted, and it's up to you to determine credibility.

Characteristics of expert information seekers

In the following sections, I will break down the steps in the information seeking process. First, however, let's look at the traits possessed by "expert information seekers"—journalists, academic librarians, and nonfiction children's book authors (Kohnen and Mertens 2019). These professionals are curious, skeptical, committed to accuracy, persistent, resourceful, and willing to change direction when necessary. They are not content with answering the basic question; rather, they explore deeper questions that arise when they encounter new information. They actively and repeatedly evaluate sources for both relevance and credibility. Keep these traits in mind as you read about the process, described next and summarized in Figure 2.2.

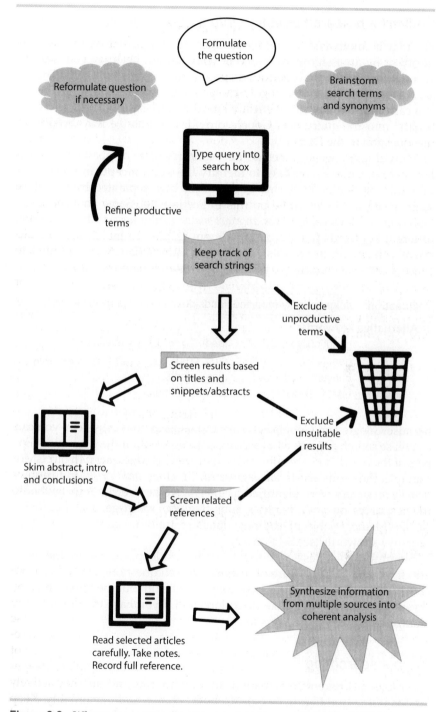

Figure 2.2   What to expect in a literature search. Concept based on UC Libraries/ UC Merced. 2020. *Begin Research Tutorial*. University of California/CC BY-NC-SA 3.0.

Convert a research question into a search query

You may be motivated to find information about a topic by your own curiosity or by an assignment in one of your classes. Let's say you want to know if genetically modified foods pose a risk to human health. The algorithms in Google and Google Scholar recognize **natural language**, so you can simply type "do genetically modified foods pose a risk to human health" into the query box. Google provides alternative ways to phrase this question in the *People also ask* section.

Natural language searches do not work in article databases, however. For these databases, you have to identify the **key concepts** in your question and then enter the terms in the query box, separated by operators such as *and*, *or*, or *not* (more on this in Section 2.4). Some databases, notably Web of Science, look for *an exact match* of the keywords. Therefore, different keywords will produce different results. To make sure you are not missing important results, make a list of alternative terms for each key concept and systematically try out different combinations of keywords.

Question: do | genetically-modified | | foods | pose a risk to | human health |

Alternative keywords:

genetically modified	food	health risk
genetically engineered	crop, crops	human health risk
conventionally bred	plant, plants	
GMO, GE, GM	animal, animals	

Because Google and Google Scholar both use a synonym system, you can use these search engines as a thesaurus. To do so, skim the titles and snippets of the first 10 to 20 results for the natural language query. In our GMO example, the terms *genetically engineered, GE crops, health risk,* and *conventionally bred crops* were identified as alternative keywords. Your textbook, the resources on your library's subject guide webpage, and PubMed's MeSH (Medical Subject Headings) database are other good sources of alternative keywords (see Section 2.4).

Skimming the titles and snippets on the results pages has another benefit: It may make you aware of related research questions that you hadn't considered. Maybe it's not the risk to human health but the impact on biodiversity that you would like to explore, and if you have the flexibility to change your topic, you can do so.

Keep a search log

After trying different combinations of search terms and different formulations of the question in Google, Google Scholar, and one or more databases, it will be hard to remember what worked and what didn't. For that reason, you may wish to keep a record of your searches.

- Which search strings yielded unproductive results?
- Which search strings yielded results that may be worth reading?

Some people prefer to keep a paper record, while others copy and paste information into an electronic document. Both methods require repeated evaluation of the search results for **relevance**; in other words, you have to ask yourself whether the content is likely to help you answer your research question. For Google searches, and especially for information found on social media, the results also have to be evaluated for **credibility**; in other words, you have to find out whether the content is scientifically accurate. Section 2.3 will cover this topic in more detail.

When evaluating search results for relevance, it is helpful to screen each result at different levels. The initial screening is the fastest and involves skimming the title and the snippet. If the result seems relevant to your topic, click the title link to open the webpage. Skim the summary (abstract), introduction, and conclusions. If the article or webpage passes this second screening, consider saving the resource in a special folder on your computer (or in the Cloud) to read carefully later.

You are probably already organizing your coursework in folders so that you can find documents quickly. For the same reason, make a folder for articles, screenshots, and other information related to your research question. Give each document an informative name so that you can identify the content later. For example, I use the following format for journal article files: **first author_ year_abbreviated title**.

Sometimes you may only have enough time to do a first-level screening of the title and snippet. Google Scholar has a convenient feature that lets you save citations to your library (see Section 2.4). Rather than having to start your search from scratch, you can go directly to your Google Scholar library and continue screening the abstract, introduction, and conclusions.

Read and take notes on the full-text article

After assembling a collection of articles on your topic, you'll read each article and take notes. Not all of the information in an article will be relevant to your research question. Therefore, skim the article first to identify the sections that contain relevant information. Then read those sections carefully, taking notes in your own words as much as possible (see Section 3.6).

Find related references using References Cited, Cited By, and Related Articles

Once you find a relevant article, Google Scholar and most of the databases make it easy to find related articles. **References cited** are found at the end of journal articles and in the more scholarly secondary sources. These are older papers that are related to the current work but have a slightly different focus. **More recent papers** that cite the current article and **papers that**

have common references can also expand your bibliography while at the same time focusing on the topic.

How to cite references will be covered in Chapter 4.

Synthesize and finalize

Most literature searches culminate in a paper or presentation in which you communicate your newfound knowledge to a particular audience of specialists or nonspecialists. Your job is to make sense of all the information you've collected. During this process of **synthesis**, you'll make connections, point out discrepancies, and support your conclusions with evidence. The goal is for your audience to understand the question, why it is important, and how you arrived at your conclusions.

2.3 Assessing the Relevance and Credibility of Web Search Results

Web search engines like Google are a convenient way to find information. In just a few seconds, Google's algorithms evaluate billions of websites based on the keywords or search query, and the results are displayed in order of relevance. Google assesses relevance based on the search terms, the perceived authoritativeness of the website, the currency of the information (depending on context), the usability of the webpages across multiple devices, your location, and your previous searches (How Google Search Works 2020). Therefore, the results that Google decides are relevant may not actually be that relevant for your scholarly search.

Furthermore, information on the web may not have been checked by any authority other than the owner of the website. It is important to evaluate the results critically, especially for controversial topics pertaining to science. The following section explains how the source features provide clues about a website's credibility.

Evaluate source features

A website's **source features** include the authors' names, their affiliations, the URL, and the date posted or published. For each search result, some information about the source features may be extracted from the URL, the title, and the snippet on the search results page. But to evaluate the source features thoroughly, you will probably need to visit the website. Here are some specific questions to ask:

- What is the author's competence in relation to the topic? Information about the author is usually located at the beginning or end of the webpage. If an author's credentials are not listed, try googling the author's name.

- What is the genre (scholarly journal article, popular magazine, blog, advocacy, or something else)? By identifying the genre, you can infer whether the author has firsthand knowledge of the topic or whether the author is covering other people's work. For scholarly research, a primary source is considered to be more authoritative than a secondary source.

- Why did the author write this text? Was this a sponsored study? Did the author report any conflicts of interest? Like infomercials on television, articles that promote a sponsored product must be approached with skepticism. Research sponsored by the National Institutes of Health (NIH) or the National Science Foundation (NSF), which are funded in the United States by tax dollars, on the other hand, can be considered trustworthy because of the thorough vetting process used to award the grants.

- Who is the publisher? When the author of a website cannot be identified, check for the publisher or sponsoring organization.

- Does the ending of the URL (.com, .edu, .gov, .org, or .net) tell you anything about the author or source? Do not assume that domain names that end with .org and .com are automatically reliable and unreliable, respectively.

- When was the information published? Is the information sufficiently current for your research topic?

If a source passes the preliminary test for credibility based on the source features, skim the text to check for other attributes of a reliable site. You can use the following list as a guide.

- Figures and data provided in tables support and complement the information in the text.

- Controversial statements and specialist knowledge are supported by references (possibly hyperlinks) to reliable sources.

- Full references are listed at the end of the document or webpage.

- Topic-specific terminology is used appropriately.

- Grammatical and spelling errors are absent.

- The language is neutral and objective, not opinionated and inflammatory.

- Advertisements are absent.

- Comments from readers are absent.

- Links to other webpages or websites work. The linked sources are reputable.

Some websites can be eliminated immediately because the source features raise a red flag. Sometimes, however, it is necessary to read the content carefully before you can determine whether a site is trustworthy.

Evaluate source content

As mentioned in the beginning of this chapter, learning involves integrating new knowledge with existing knowledge. Fortunately, the same skill set required for learning is also used in the evaluation of source content, as illustrated by the following two studies. In the first study, juniors and seniors in a Finnish high school were asked to write an essay about internet censorship (Killi and Leu 2019). Working in pairs, students first discussed the topic and constructed a mind map based on their prior knowledge (10-15 minutes). Then they were given 30 minutes to search for additional information on the web and to revise their mind maps based on their online reading. Finally, each pair spent 45 minutes writing a joint essay based on their online research.

In the second study, undergraduates at a public university in the Midwest were instructed to use seven websites returned in a Google search to answer the question "What caused the eruption of Mt. St. Helens volcano?" (Goldman et al. 2012). The students were pre-screened to make sure they had minimal knowledge of plate tectonics and volcanic eruptions. The students were instructed to "think out loud" as they read online and navigated among websites, verbalizing what they were focused on, what the information made them think about, and why they decided to read what they did. The researching/thinking aloud part of the experiment was videotaped so that later the researchers could relate students' comments to their actions. The students had up to 50 minutes to research the topic online and 40 minutes to write their essays.

Findings from both studies indicated that better students:

- Repeatedly **compared and contrasted** new information with what they already knew. They asked themselves if the new information made sense (monitored prior knowledge) and tried to extract meaning out of what they were reading (practiced self-explanation). The process of integrating new with existing knowledge had an added benefit: it enabled better students to distinguish information that was task-relevant and scientifically accurate from information that was not.
- **Displayed persistence**. When they didn't understand a sentence, they continued to read until they understood.
- Navigated to a different website because they (1) didn't find the specific information they were looking for or (2) wanted to check if the new information was supported on another website. Better

students used navigation strategies that were **goal-oriented** and included **purposeful fact-checking**.

- Spent **more time reading new information** than information that was familiar. New information has to be read slowly to give our brains time to process it. Familiar information, on the other hand, can be skimmed because reviewing requires less processing than initial learning.
- Incorporated **more new information** than prior knowledge into their essays.

Notice how the characteristics of better students align with those of the expert information seekers described in Section 2.2.

Conversely, findings from the two studies revealed that less capable students:

- Tended not to compare and contrast new information with what they already knew.
- Tended not to check information for consistency across multiple websites.
- Spent little effort trying to process new information.
- Skimmed websites superficially for keywords and, if they didn't find the right keywords, navigated to a different website.
- Viewed online reading as a task that had to be checked off a list, rather than an opportunity to acquire new knowledge.
- Spent the same amount of time reading inaccurate and accurate information. In other words, less capable students spent less time monitoring their prior knowledge and evaluating new information in the context of that knowledge.
- Incorporated fewer accurate core concepts into their essays. The less capable students relied more on prior knowledge and knowledge activated during reading than new knowledge gained from online reading. They were content with finding information that confirmed what they already knew or believed, instead of engaging actively with new material.

Here is the takeaway message for evaluating online sources: **be skeptical and repeatedly check your facts.** Does the new information align with your previous knowledge about the topic? Is the new information in agreement across numerous trustworthy websites and print publications? Do authors or publishers of the website have the expertise to write about this topic? Are they motivated by something other than money? Paying attention to the habits of both the better students and the less capable students can help us become smarter consumers of information. For additional resources for checking credibility, please see Box 2.1.

BOX 2.1 **Resources for checking credibility**

❏ The CRAAP Test. This test is described on many library websites.

❏ Fake news. These fact-checking sites call out fake news: AllSides (https://www.allsides.com/unbiased-balanced-news), FactCheck.org (https://www.factcheck.org/), Hoaxy (https://hoaxy.iuni.iu.edu/), and Snopes (https://www.snopes.com/). More sites available from Benedictine University (https://researchguides.ben.edu/fake-news).

❏ Ulrich's Periodicals Directory. Your academic library may subscribe to this database. Type the journal name into the search box. The referee icon indicates that the journal is peer reviewed.

❏ Retraction Watch (https://retractionwatch.com/). This organization tracks scholarly articles that were retracted before or after publication. Articles are retracted (withdrawn) by authors or journal editors for various reasons that may include the discovery of a major error in the manuscript, poor experimental design, falsified or fabricated data, misconduct after the fact, or the inability of other researchers to reproduce the results. The database is searchable.

❏ Quality Checklists for Health Professions blogs and podcasts (DOI:10.15200/winn.144720.08769)

2.4 Strategic Use of Search Engines and Databases

Choosing an article database

Article databases contain a pre-screened collection of journal articles, books, conference papers, government publications, and other types of documents. Some of the biology databases you may have access to include Agricola, BioOne, Biological Abstracts, BioMed Central, BIOSIS Citation Index, GEOBASE, JSTOR, PsycINFO, PubMed, and Web of Science. Each database contains a slightly different collection of scholarly information. For example, if your research topic were in the field of ecology, a search in Web of Science would be more productive than a search in PubMed, which specializes in the biomedical literature. If your research project requires a comprehensive literature review, however, it is important to search multiple databases.

If you are just starting out, click the subject guide link on your library's website, select the subject that is most relevant to your topic, and choose one of the recommended databases. Schedule an appointment with your subject librarian to learn how to get the most out of these resources.

Keyword searches

Database algorithms respond to one **keyword** or **a string of keywords**, but not natural language queries. Effective keyword strings are neither too broad nor too narrow in scope. Strings that are too broad will retrieve an unmanageable number of articles that, for the most part, are not relevant to your topic. On the other hand, keyword strings that are too specific may not give you any results. As mentioned in Section 2.2, you can increase your chances of finding relevant sources by trying different keyword combinations in the query box. In some databases, different word endings (photosynthe*sis* versus photosynthe*tic*), abbreviations (*HIV* for *human immunodeficiency virus*), and alternative spellings (American versus British English) may also produce different results. Vague terms like *effect of* and *relationship between* should not be used as keywords.

Keyword strings can be refined if the search results are missing the mark. Here are some suggestions.

Connect keywords with Boolean operators such as *and*, *or*, or *not*.

- When the word *and* is used between keywords, the references must have both words present. This connector is a good way to *limit* your search.

- When the word *or* is used, the references must have at least one of the search terms. This connector is a good way to *expand* your search.

- When the word *not* is used, the references should not contain that particular keyword. This connector is another way to *limit* your search.

Note: In Google and Google Scholar, these operators must be written in ALL CAPS to distinguish them from regular words.

Another way to *limit* your search is to search for the exact phrase. To do so, enclose the phrase in double quotation marks. As an example, the term *migrating seabirds* produced over 29,000 results in Google Scholar, whereas *"migrating seabirds"* produced only 378. (Another search may not result in the same numbers, but the difference would likely be of the same magnitude.)

When keywords have multiple endings, you can use truncation to *expand* your database search. For example, many words related to the concept of temperature begin with *therm*, such as thermoregulation, thermoregulatory, thermy, and thermal. Rather than writing a lengthy search string containing all of these terms, simply type *therm**. The asterisk is the truncation symbol used in many databases, but you should check the Help menu for confirmation.

Sample search

To illustrate how Google, Google Scholar, and two different databases (Web of Science and PubMed) can be used interactively to find information, let's explore the topic *how climate change affects food availability for migrating seabirds*. Although secondary references may provide background information, the goal of our search is to find primary references on this topic.

Google When I typed this search query into Google on July 13, 2020, more than 1.4 million results were returned; most of them were *not* primary references. **How can you tell if a result is a primary or secondary source?** To answer this question, let's look at a section of the first page of search results (Figure 2.3).

Notice that each result has a URL, a title, and a snippet. Just from skimming these three features (and with a bit of experience), you can tell that Audubon and USDA are secondary sources. Clicking on each of these links and skimming the content of these websites confirms that they are intended for a general audience. They do *not* possess the attributes of primary references; in particular, the authors do *not*

- Provide details about procedures or data
- Use jargon
- Cite sources to support their assertions or list full references at the end.

On the other hand, the web addresses for NCBI, Frontiers in, and JSTOR indicate that these results link to article databases or scholarly journals. These are primary references that may be relevant to the research question. Recent publications (within the past 12 months) may not be accessible without a subscription, which is the problem you run into when you click the JSTOR link. If your library has a JSTOR subscription, you will have to log in through your university account to gain access to this article.

The **Frontiers in** link provides access to a potentially useful secondary reference—a review article. **Review articles** summarize the information in the most important primary references published during the review period. Review articles not only provide a broader overview of the topic than any individual journal article, they are also an excellent source of primary references.

Google Scholar An efficient way to eliminate ads and secondary references from a web search is to use Google Scholar (https://scholar.google.com/). Google Scholar's algorithm uses information provided by authors and publishers to limit searches to scholarly websites across many disciplines. Search results include not only journal articles but also material from the websites of universities, scientific research groups, and professional societies; conference proceedings; court opinions and patents; and preprint archives. **Preprints** are manuscripts that authors upload to a server (for example, bioRxiv.org or medRxiv.org) to disseminate current information before peer review and publication. Google Scholar's search results do not include articles in the popular press, book reviews, or editorials.

Returning to our example, clicking the "Scholarly articles" link near the top of the Google search results page takes you to the Google Scholar

Google

how climate change affects food availability for migrating seabirds   ✕   🎤   🔍

🔍 All   📰 News   🖼 Images   ▶ Videos   🛒 Shopping   ⋮ More   Settings   Tools

About 1,430,000 results (0.34 seconds) —— Number of results

Advertisements are labeled ——

Ad cropscience.bayer.com/climate_change/agriculture ▾
Climate Change & Agriculture - Shaping The Future Of Ag
See How Bayer is Helping Farmers Manage **Changes** in Weather with **Climate**-Smart Solutions.
We're Committed to Shaping Agriculture & Finding Solutions in Addressing **Climate Change**.
Sustainability Commitment. Shaping Agriculture. Helping Farmers Thrive.

Link to Google Scholar ——

Scholarly articles for **how climate change affects food availability for migrating seabirds**
... conservation of **seabirds** facing global **climate change**: ... - Grémillet - Cited by 235
... signals drive breeding phenology of three **seabird** ... - Frederiksen - Cited by 227
Seabirds and **climate**: knowledge, pitfalls, and ... - Oro - Cited by 30

National Audubon Society ——

www.audubon.org › magazine › september-october-2014 ▾
How Climate Change is Sinking Seabirds | Audubon
Even small shifts in temperature or chemistry or salinity can have cascading **effects**. Warming is
reducing the basic amount of **food** produced by oceans. It's also **changing** the proportions of
what lives in the sea—the ratio of sardines to anchovies, for example.
You've visited this page 2 times. Last visit: 7/12/20

PubMed Central archive at the National Library of Medicine ——

www.ncbi.nlm.nih.gov › pmc › articles › PMC2781852 ▾
The impacts of climate change on the annual cycles of birds
Migration and reproduction of many avian species are controlled by endogenous .... **Climate
change** is causing mismatches in food supplies snow cover and other ... of species of **seabirds**
in east Antarctica are delaying arrival and egg laying. ...
by C Carey - 2009 - Cited by 164 - Related articles

Frontiers in Ecology and Evolution journal ——

www.frontiersin.org › articles › fevo.2014.00079 › full ▾
Seabirds and climate: knowledge, pitfalls, and ... - Frontiers
Dec 8, 2014 - **Climate Change** and Marine Top Predators View all 15 Articles ... (at a spatial
mesoscale) to trans-equatorial **migrating seabirds** that travel large ... mechanisms, **food
availability** and its effects on seabird population dynamics ...
by D Oro - 2014 - Cited by 30 - Related articles
Climate and its influence on ... · The Mechanisms Linking ... · Predicting the Future ...
You visited this page on 7/12/20.

United States Department of Agriculture ——

www.fs.usda.gov › ccrc › topics › wildlife › birds ▾
The Effects of Climate Change on Terrestrial Birds of North ...
Research on birds has shown that **climate change affects** birds both directly ... **food**
requirements of the young coincides with the maximum **food availability** (14). ... to climate and
is advancing in most regions migratory bird species are in some ...
You visited this page on 7/12/20.

JSTOR is a digital library ——

www.jstor.org › stable
Seabirds and climate change - jstor
May 21, 2020 - **Climate affects food availability** to planktivorous ... understanding of **climate
change effects** on **seabirds** at the global scale, however, additional low- ... Ocean warming and
lengthened growing season allow delay of migration. —— Includes synonyms
by WJ Sydeman - 2012 - Cited by 89 - Related articles   and variations of
     search terms

Figure 2.3   Google search results page for *how climate change affects food availability for migrating seabirds.*

results page. As shown in Figure 2.4, Google Scholar returned "only" 16,200 results. Google Scholar's Advanced Search provides basic filters for reducing the results to a more manageable number (Figure 2.4). Search results can be filtered by date, author, publication, and characteristics of the search string. However, you cannot change the criteria Google Scholar uses to rank the relevance of the results.

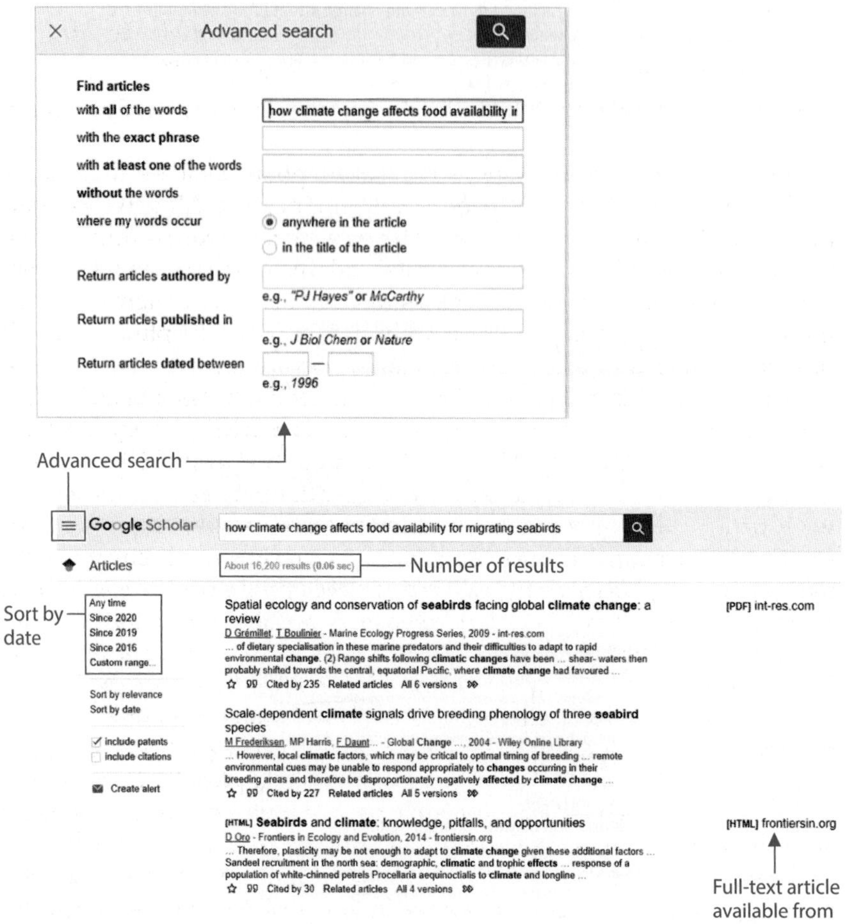

Figure 2.4 Google Scholar search results page for *how climate change affects food availability for migrating seabirds*. Search results can be limited by date, author, publication, and where in the article the search terms occur. Full-text articles can be retrieved by clicking links to the right of the results.

Google Scholar also has a number of useful citation features. Below the snippet, you will see:

☆ **Save the citation** to your Google Scholar library for additional screening later.

🔟 **Copy the full reference** to clipboard to paste into a document. Of the styles listed, American Psychological Association (APA) is closest to the Council of Science Editors (CSE) style that will be described in detail in Chapter 4. The reference can also be copied into BibTex, EndNote, RefMan, or RefWorks. These are reference management programs that organize references and facilitate the formatting of in-text and full references.

Cited by and **Related articles** link to related references (see Section 2.2)

Web of Science   Web of Science is a subscription database that covers many subject areas in the life sciences and engineering. All of the journals in this database are peer-reviewed. The peer review process provides assurance that the information is accurate, and this is one reason why scientists generally prefer to use databases rather than search engines to seek information. If your library has a Web of Science subscription, you will need to log in to your university account to gain access.

Database algorithms recognize keywords, not natural language queries. In our example on *how climate change affects food availability for migrating seabirds*, the key concepts would be **climate change**, **food availability**, and **migrating seabirds**. We would use the *and* operator between words because we want all three concepts to be included in the results (see "Keyword searches" in Section 2.4). Surprisingly, only six results were returned with the search string *climate change and food availability and migrating and seabirds*, compared to over 16,000 in Google Scholar. A possible reason for the small number is that Web of Science's algorithm looks for *an exact match* of the keywords in *just the title and the abstract*. On the other hand, Google Scholar searches the entire document, and it automatically uses a synonym system that recognizes related terms. Keep in mind that in Google Scholar, the operators AND, OR, and NOT must be capitalized in keyword search strings.

To refine our Web of Science search, we can revise our query by using alternative keywords (see Section 2.2) and truncation (see Section 2.4). Truncation alone (*climat* and food and migrat* and seabirds*) increased the number of search results to 55. Other unique references were located by replacing the term *migrating seabirds* with *shearwaters, auklets,* or *petrels* (or their corresponding Latin names).

PubMed   PubMed is *the* most recommended database for researchers in medicine who require advanced search functions. Like Google

Scholar, PubMed is available free of charge to the public. Its advanced search features make it possible to limit searches by author, publication, and date. Beyond these basic filters, PubMed provides a variety of options to retrieve only certain formats (full text, free full text, or abstract), type of article (clinical trial, review, clinical conference, comparative study, government publication, etc.), language, and content (journal group, research topic, humans or animals, gender, and age). PubMed also links to other National Center for Biotechnology Information (NCBI) resources, such as GenBank, BLAST, and taxonomy databases (PubMed Help 2020).

Another feature that makes PubMed so powerful is its search algorithm, which is based on concept recognition, not letters or words. Every document indexed for PubMed has been read by experts, who tag the document with **controlled vocabulary** (MeSH) that accurately describes the paper's content. MeSH solves the problem of ambiguity concerning scientific and popular names of organisms, synonyms, and variations in British and American spelling. MeSH is more structured than the synonym system used by search engines, which means that the database will return a smaller number of more relevant results.

Let's use an example in the biomedical field to illustrate the advantages of a PubMed search. Say you would like to know if there is a *cure for color blindness*. We suspect that there may be a medical term for *color blindness*, so let's look it up in the M̲edical S̲ubject H̲eadings database. On the PubMed homepage (https://pubmed.ncbi.nlm.nih.gov/), we'll click **Explore | MeSH Database**, type *color blindness* into the query box and then choose the subheading *inherited color blindness* for this example. The search result tells us that the controlled vocabulary for this term is *color vision defects* (Figure 2.5). To search for articles in the PubMed database with this term, simply click the **Add to search builder** button on the right panel and then **Search PubMed**. The search returned 4006 results (in July 2020).

Since we would like to know if there is a cure for color blindness, we can limit the search by clicking the **therapy** subheading, **Add to search builder**, and **Search PubMed**. This search returned 193 results. There were two review articles among the first 20 results. These would probably give a good overview of the current state of knowledge and indicate which of the other articles might be relevant to our topic.

Keywords can be combined with MeSH. For example, *contact lenses* and *gene therapy* are terms we may not have considered initially but may have become aware of from reading some of the search results titles. To pursue these two subtopics, add *color vision defects* to the PubMed Search Builder as before and then type *AND contact lenses*. This search string reduces the number of results to 37. Alternatively, *color vision defects [MeSH] AND gene therapy* yielded 110 results.

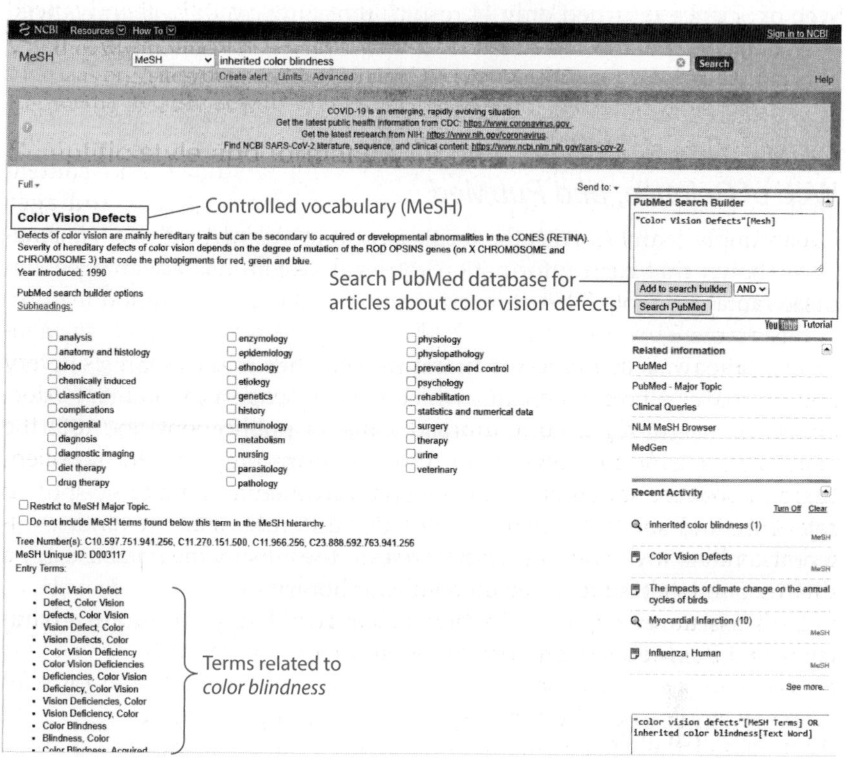

Figure 2.5 PubMed can be searched with keywords or MeSH or a combination of the two.

The other way (besides MeSH) to search the PubMed database is to use keywords. Because every article in the database has been tagged with controlled vocabulary, the search string *color blindness and gene therapy* not only yields results with those exact keywords in the title and abstract, it also retrieves articles with the words *colour, genetic, color-blindness, defects, vision, gene replacement, gene editing,* and *gene augmentation therapy.* This feature of PubMed saves you time brainstorming synonyms or entering long search strings for every possible variation of the keywords.

How do the search results for *color vision defects and gene therapy* in PubMed compare with the results in the search engines and Web of Science (in July 2020)? Google returned over 11 million results, which were not limited to scholarly articles. Google Scholar returned over 43,000 results—many posted on academic social media sites (ResearchGate and Academia.edu), company websites (.com), university websites (.edu), professional societies (.org), and other locations not indexed by databases.

Web of Science returned only 14 results; replacing *color vision defects* with *color blindness* returned 54 results, demonstrating the importance of using alternative keyword combinations in Web of Science searches.

Comparison of Google, Google Scholar, Web of Science, and PubMed

The example searches with the terms *migrating seabirds* and *color blindness* have shown that each information retrieval system has advantages and disadvantages. Table 2.1 summarizes some of the most common features of these systems.

The **takeaway message for literature searches** is that search engines and databases all return results, which vary in scope and quality. Google Scholar is useful for finding a broad range of publications and offering alternative means to access them. Databases are preferred for scholarly searches because the content has been pre-screened for both topic-specific relevance and scientific accuracy. Because different databases index different content, if you are not finding relevant results in one database, try a different one that is recommended on your library's subject-guide website. Or seek assistance from your reference librarian. Use your resources strategically for an efficient and productive search.

2.5 Social Media

Social media are "forms of electronic communication through which users create online communities to share information, ideas, personal messages, and other content" (Merriam-Webster online dictionary). Examples of social media that have science content are forums (discussion groups), blogs, wikis, videos, and social networking sites like Facebook, Twitter, Instagram, LinkedIn, ResearchGate, Academia.edu, and others. Communication in these online communities is not one-to-one, but many-to-many, which has both advantages and disadvantages.

The purpose of this section is to make you aware of how educators, researchers, and healthcare professionals use social media to disseminate scientific information. However, because social media posts rarely discuss the specifics of primary references, your primary source of scholarly information should continue to be article databases. Furthermore, information obtained from social media requires extra fact-checking work from you to assess scientific accuracy because anyone can post information, fact or not, on these sites.

Social media platforms that are popular among scientists

Education   Social media are often used in your formal education without you even noticing. Besides the social media links that you may have

TABLE 2.1 Comparison of selected search engine and database features				
Feature	Google	Google Scholar	PubMed	Web of Science
SEARCH TERMS				
Natural language search is possible	✓	✓		
Double quotation marks around keywords enable search for exact phrases	✓	✓	✓	✓
Uses synonym matching or controlled vocabulary	✓	✓	✓	
Recognizes different word endings	✓	✓	✓	
Recognizes truncation symbols such as asterisks			✓	✓
SEARCH RESULTS				
Keyword search limited to title and abstract			✓	✓
Search results include secondary references	✓			
Search results include advertisements	✓			
Social media search is possible (type @ or # followed by keywords)	✓	✓		
Content is limited to scholarly publications		✓	✓	✓
Full-text articles may be accessed through scientists' homepages and course websites	?	✓		
CREDIBILITY				
Most or all results are peer reviewed			✓	✓
LINKS TO SIMILAR ARTICLES				
Search results have links to references cited, cited by, and related articles*		✓	✓	✓
SEARCH RESULTS REFINEMENT				
By year	✓	✓	✓	✓
By author	✓	✓	✓	✓
By document type	✓	✓	✓	✓
Using advanced options			✓	✓
SAVING REFERENCES				
Full reference of journal article can be copied to clipboard		✓	✓	
Full reference of journal article can be copied to reference manager (e.g., Mendeley, Zotero, RefWorks, or EndNote)		✓	✓	✓

* JSTOR's Text Analyzer is another tool for finding related articles. See Bibliography.

Source: Based on input from Kathleen McQuiston, Research Services Librarian, Library and Information Technology, Bucknell University (2016 Oct 23), updated for current database and search engine websites.

accessed during the college application process, individual departments may have Facebook pages and a presence on Twitter to showcase faculty research and community outreach. These links (along with the department website) are good ways to learn about the research interests of individual faculty members, so that you know who to contact in case you are interested in doing research.

In the classroom, your instructors may show **YouTube** or **Vimeo** videos in their lectures to generate interest in a topic or to explain a concept that would be hard to understand just by reading about it. Because anyone can post videos online, the identifying information (affiliations and sponsors) should be evaluated to ensure the source is credible.

Most of us have used **Wikipedia**, the online encyclopedia, to get background information about a topic. A **wiki** is a website that allows members of a specific community to collaborate on content. Most of Wikipedia's volunteer editors take their responsibility to provide quality content seriously, although mistakes sometimes happen. One way to gauge the quality of a Wikipedia article is to look at the references at the end of the page. More references mean more opportunities for fact-checking.

TED Talks (https://www.ted.com/talks) are short presentations by "some of the world's greatest thinkers, leaders and teachers" about a wide range of topics. Originally, the talks were focused on the intersection of technology, entertainment, and design, but now there are more than 3500 searchable TED talks, nearly one-third on science topics. Presenters and their work are vetted by the organization's science curator to make sure the information is accurate and up-to-date.

Professional networking sites   **LinkedIn** is a networking site for professionals in all fields. You build your network by inviting other professionals to join yours and to accept invitations from others to join theirs. The connections are primarily useful if you are actively looking for a job. If you choose to join LinkedIn, keep your profile up-to-date, as employers may use LinkedIn to pre-screen applicants. Furthermore, make sure your online presence is "clean" because personal websites and blogs may also be accessible through LinkedIn. Remove any posts containing partying pictures, offensive language, and political comments, as these may suggest to employers that you are not a good fit for the position.

ResearchGate and **Academia.edu** are professional networking sites specifically for science, technology, engineering, and mathematics (STEM) researchers. (Students are generally not eligible to join unless they are actively doing research.) Members can create a profile, promote their work, network, collaborate, upload their publications for sharing and review, and obtain statistics on these publications. You may have noticed these sites on the Google Scholar search results pages, where full-text articles are sometimes linked through ResearchGate or Academia.edu instead of a journal publisher's website.

Twitter   Twitter has emerged as a leading platform for the dissemination of scientific knowledge. Although less than 5% of research papers were shared or mentioned on Facebook, 21% received at least one tweet (Diaz-Faes et al. 2019). Furthermore, 10% of the 1.4 million papers indexed by PubMed and Web of Science have been tweeted (Mohammadi et al. 2018).

What kind of information do scientists share on Twitter? Scientists tweet breaking news, such as a new publication or real-time information at a conference (Mohammadi et al. 2018). The links to social media on journal websites make it easy to share publications with other scientists (Figure 2.6). However, scientists also tweet about everyday topics of interest. For example, they share videos and pictures of their research organisms and informal publications such as magazine and newspaper articles and blog posts, and they also seek advice through Twitter (L. Naughton, personal communication, August 23, 2020).

In a wonderful example of serendipity, the posting of an image to Twitter resulted in a reexamination of a misidentified plant specimen. Within hours of posting the image, Dr. Chris Martine received a reply from an observant botanist in the community, who suggested that the specimen that had been identified as *Heuchera americana* looked more like *Heuchera alba* (Schuette et al. 2018). After carefully comparing the specimen with those in herbarium collections, the Martine research group was able to confirm the specimen's identity as *H. alba* and set the record straight.

Over the past decade, more scientists have become interested in **science communication**, which means communicating science to a general audience, not just an audience of other scientists. The hashtag #scicomm links to conversations about science communication on Twitter and Instagram. In addition, #scicomm toolkits can be found on the American Association for the Advancement of Science (AAAS) and the American Geophysical Union (AGU) websites.

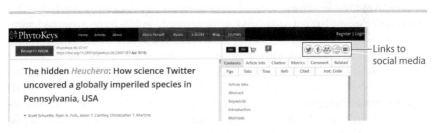

Figure 2.6   Academic journal websites provide links to social media and email to facilitate sharing of articles.

Scientific organizations such as NASA, National Geographic, The Smithsonian Institution, and many others also have a presence on Twitter. At the time of this writing during the COVID-19 pandemic, the Centers for Disease Control and Prevention (CDC) provides "daily credible health & safety updates" through Twitter. All of these organizations provide information that can be easily understood by a general audience. If you are conducting a scholarly literature search, the links in the tweets may help you acquire background information needed to understand the primary literature.

How do you find information on Twitter? If you are new to Twitter, just type a topic into the search box. The results can be filtered by Top, Latest, People, and media type (photos and videos). Twitter also has an advanced search feature that can be accessed by clicking the ellipse (...) just below the search box or **Advanced search** on the righthand panel below **Search filters**. As in Google Scholar and the databases, the results can be limited by setting restrictions on the search terms and dates. In addition, filters can be applied to accounts, tweets, replies, likes, and retweets.

Because social media are based on relationships, most Twitter users follow a specific account (@username) or search for keywords preceded by hashtags (#keywords). Many scientists organize the accounts and topics they like to follow using Twitter Lists. A **List** is "a curated group of Twitter accounts" (Twitter, Inc. 2020). Private lists are only accessible to the account holder; public lists can be followed by anyone. Based on a user's activity, Twitter may suggest other Lists that might be of interest with the prompt "Discover new Lists." Other tips for using Science Twitter are offered by James (2020).

Twitter has emerged as a leading platform for two other areas of interest to scientists. One area is informal peer review of published articles. For example, #arseniclife became a case study on how the controversial findings in a NASA research article were debated on Twitter (Yeo et al. 2017). Because of the skepticism scientists voiced on Twitter, another research group set out to repeat the experiment. After 14 months, the publication was retracted because the results could not be replicated.

Another area in which Twitter has emerged as a front-runner is the celebration of scientific interests and achievements of underrepresented communities (for example, #BlackBotanistsWeek, @SACNAS, #womeninSTEM, among others). Like LinkedIn, Twitter makes it possible to follow or connect with scientists who inspire you, whether or not you are looking for a job.

Other online resources with science content   Blogs (short for "weblogs") are informal, comparatively short posts that are updated regularly. The posts are listed in order of currency, with the most recent first. Science bloggers communicate science to a non-specialist audience and may use their experience to put the science in context. Because of these characteristics, blogs have the potential to be good secondary sources in your literature search.

Programming sites. The programming language R is increasingly being used by biologists for statistics, data processing, and data visualization, especially in ecology and evolution, bioinformatics, and "big data." Two useful sites for programmers are **Stack Overflow** and **GitHub**. Stack Overflow is a Q&A site where people post questions about their code, and other people propose answers. Answers are evaluated by members of the community with up or down votes. Stack Overflow is usually used for quick debugging questions and for finding code to reuse and adapt to a new project. **GitHub** is reputed to be the world's largest code-hosting website. People post complete projects to allow other researchers to reproduce their results, to draw attention to their work, and to allow others to build on their code. GitHub is also a good source of more advanced statistical models (C. Kelling and B. Knisely, personal communication, September 5, 2020).

Reddit is a social news and media aggregation website where registered users can submit content and vote posts up or down. Reddit has subsections called subreddits, where students can share information and get answers to questions (https://www.reddit.com/r/science/). Other communities discuss biology, botany, microbiology, neuroscience, and other topics. This is not a source of primary references; it is an informal Q&A venue. Because Reddit posts are not reviewed for accuracy, make sure you confirm the information using reliable sources.

Scientists and others may consult many online sources to stay current on topics of interest. To avoid having to check multiple websites for new content, **RSS (really simple syndication or rich site summary) feeds** make the content come to you. To subscribe to an RSS feed, click the orange feed icon or the "subscribe to RSS feed" button on your favorite websites. Then select a feed reader on the web; if you would like to receive the feeds on your cell phone, select a reader specific for mobile devices. A feed reader aggregates all of your feeds in one place to make them easy for you to read. The websites for scholarly databases such as Web of Science and PubMed also offer subscriptions to RSS feeds.

Finally, in mainland China, where Twitter and Facebook are unavailable, scientists use **WeChat** to discuss their recent publications, share information, and invite collaboration (*Nature Cell Biology* Editorial, 2018).

Healthcare advice on social media—room for improvement

Social media platforms also offer a way for medical professionals to communicate directly with the public. As pointed out by Barton and Merolli (2019), there is a great need for medical researchers to translate the knowledge in journal articles into usable information for nonspecialists using, for example, blogs, infographics, videos, and podcasts. This content can then be disseminated through a variety of social media platforms. James (2020) describes how COVID-19 researchers, scientists in other fields, physicians, journalists, public policy leaders, and others have exchanged information on Twitter during the COVID-19 pandemic. Their conversations include specific questions about a journal article, requests for hospital protocols or lab supplies, correction of misinformation, and other topics of interest to both specialists and a general audience.

As the worldwide response to the COVID-19 pandemic has made abundantly clear, getting timely and accurate information to the general population is critical for controlling the outbreak. What types of media will reach the greatest number of people? A Pew Research Center survey revealed that between October 2019 and June 2020, 25% of U.S. adults got their news from news websites or apps, 18% from social media, 16% from cable or local TV, 13% from network TV, 8% from radio, and 3% from print sources. However, those who relied on social media for news tended to be less aware and less knowledgeable about current political events and national news about the coronavirus than those in the other groups (Pew Research Center, June 2020). Social media users were also more likely than the other groups to have been exposed to conspiracy theories about the COVID-19 pandemic.

These statistics bear out the biggest problem with social media: the lack of quality control. Uninformed users have the same right to comment as experts. These ill-informed comments are propagated instantly to thousands of friends and followers by a simple like or tweet. Because human nature is to like what is familiar, misinformation may spread as fast as, if not faster than, accurate information (Gierth and Bromme 2020). As pointed out by Lima et al. (2020), "we are living not just in a pandemic, but also in an 'infodemic' where fake news is becoming more common."

What can you believe on social media? A good approach is to be skeptical: don't believe anything you read until you confirm through other (reliable) sources that the information is reliable. Build up a small network of trusted users who keep up with current events. Just as you would evaluate website source features (see Section 2.3), find out which specific people or organizations are behind the usernames. There is a big difference in the trustworthiness of information posted by @NatGeo (National Geographic) versus that posted by individual users whose authority to comment on issues related to science and health may be questionable.

When *you* use social media, please follow The Golden Rule: treat others the way you would like to be treated. Be professional and respectful in all posts. Participate only when you have the expertise and only when your contribution adds value. Read and follow your organization's policy on social media, which typically addresses issues such as no tolerance for illegal or criminal acts; protecting trade secrets and intellectual property; keeping confidential information private; and treating faculty, staff, and students with respect.

Check your Understanding

1. Search for information on a topic of interest in Google. Some possible topics to explore include climate change, emerging infectious diseases, endangered species, evolution, genetically modified foods, stem cell research, stress and aging, synthetic biology, and vaccines. For the first 20 results, determine which are primary and which are secondary references. Be prepared to explain how you know.

2. Explain how and when you would use primary versus secondary references for a research project.

3. Describe how expert information seekers would go about evaluating a website's trustworthiness.

4. Search for information on a controversial topic in science (see No. 1 for ideas). Apply the strategies for evaluating web sources, described in Section 2.3, to selected search results. Write a summary of your findings, which includes a reflection on the effectiveness of the strategies you used. Materials for a sample activity "Should I get a flu shot?" are available at **macmillanlearning.com/knisely6e**.

5. Search a topic of interest in Google, Google Scholar, and a database appropriate for the topic. Compare and contrast the results in terms of number, relevance to topic, and credibility. What are the advantages and disadvantages of each information retrieval system for returning results on this particular topic?

6. Search a hot topic in science on social media. Compare the numbers of posts related to journal articles across different platforms. Do your findings agree with those of Diaz-Faes et al. (2019) that more journal articles are shared or mentioned on Twitter than Facebook?

7. Use social media to find a post that mentions a journal article. Then search for the full-text journal article on Google Scholar or in an appropriate database. Were there any restrictions on your ability to retrieve the full-text article (for example, subscription-only)?

8. Follow an active, inspiring STEM researcher on the social media platform of your choice. Keep a record of the current events and publications you learned about through this researcher over the course of the semester.

9. Identify an "influencer" on the social media platform of your choice. Over the course of the semester, keep a record of information posted by this influencer, which could be useful to you professionally (not privately).

10. Search for information about a controversial topic on Science Twitter (see No. 1 for ideas). Identify posts for specialists by specialists (technical posts) and other posts intended for a general audience (science communication). How do these posts differ with regard to language, readability, and content? What are the benefits and drawbacks of each form of communication for learning about this topic? How can you tell whether the information provided is accurate?

Go to **macmillanlearning.com/knisely6e**
and select **"Student Site"** to access
samples, template files, and tutorial videos

## Chapter 3

Reading and Writing About Science

Objectives

3.1 Understand the IMRD format of scientific papers

3.2 Learn how to read scientific papers to locate specific information

3.3 Understand how audience and purpose determine writing style and content (genre)

3.4 Learn how communicating science to a general audience differs from communicating science to other scientists

3.5 Learn how to read your textbook

3.6 Learn how to take notes on your reading and in your classes

Reading and writing are two sides of the same coin. We read to acquire knowledge and write to disseminate it. Acquiring knowledge in biology is not just about memorizing facts and practicing lab techniques. Knowledge acquisition is a lifelong process that involves mastering the basics, applying basic information to new situations and problems, and reviewing and reciting the information at regular intervals until it becomes second nature. Disseminating knowledge through written and oral communications requires a certain mastery of the subject matter. Your understanding must then be translated into words that communicate knowledge clearly and concisely to your audience.

3.1 Types of Communications

Writing about science takes many forms. As an undergraduate biology major, you will be asked to write laboratory reports, answer essay questions on exams, summarize information from journal articles, and do literature surveys on topics of interest. Third- and fourth-year college students may write research proposals and honors theses and present their work at poster sessions and other venues. Graduate students typically write master's theses and doctoral dissertations and present talks about their research at national and international conferences. Professors write lectures, letters of recommendation for students, journal articles, and grant proposals. They also review journal articles and grant proposals submitted by their colleagues. In business and industry, common forms of writing include progress reports, product descriptions, operating manuals, and sales and marketing material. Medical writers research and prepare various kinds of documents and educational materials for healthcare professionals, pharmaceutical companies, and regulatory agencies. Journalists write about science for a broad, non-specialist audience.

Many professionals are using social media and community outreach initiatives to share their passion for science with a wider audience. Communicating science to nonscientists requires a different way of thinking, because the purpose of science news is to celebrate scientific findings, rather than validate them (Fahnestock 1986). Besides raising awareness of careers in STEM fields, science communication (#SciComm) can build trust in science and increase science literacy to enable better decision-making on the part of the public. Learning how to explain scientific concepts and jargon in plain language is also good practice for paraphrasing (see Section 4.2). Finally, the creativity and technical know-how needed to put together a multimedia presentation for a general audience are good skills to put on a resume.

The next two sections explain how research papers are structured and how to read them to find specific information. In Section 3.4, we will compare a research paper with a science news article on the same topic. You will see that writing for a general audience is not just about translating jargon, but telling a story that captures the imagination.

3.2 IMRD Format

Research papers, or journal articles, are descriptions of how the scientific method was used to study a problem. Journal articles follow a standard format that is sometimes called the IMRD format. **IMRD** is an abbreviation of the core sections of a scientific paper—Introduction, Materials and Methods, Results, and Discussion. This format is very convenient, because it allows busy people to skim a paper to see if it contains the information they seek, and then spend more time reading the details.

In addition to the core sections, almost all scientific papers also have:

- A title
- A list of authors and their affiliations
- An abstract
- Acknowledgments and
- A references section

The **Title** is a **short, informative description of the essence of the paper**. It should contain the fewest number of words that accurately convey the content. Readers use the title to determine their initial interest in the paper.

Only the names of **people who played an active role** in designing the experiment, carrying it out, and analyzing the data appear in the **List of Authors**.

The **Abstract** is a **summary of the entire paper** in 250 words or less. It contains (1) an introduction (scope and purpose), (2) a short description of the methods, (3) results, and (4) conclusions. There are no literature citations or references to figures in the abstract. If the title sounds promising, readers will use the abstract to determine if they are interested in reading the full text.

The **Introduction** concisely states what motivated the study, how it fits into the existing body of knowledge, and the objectives of the work. The introduction consists of two primary parts:

1. **Background or historical perspective on the topic**. Primary journal articles and review articles, rather than textbooks and newspaper articles, are cited to provide the reader with direct access to the original work. Inconsistencies, unanswered questions, or new questions that resulted from previous work set the stage for the present study.

2. **Statement of objectives of the work**. What were the goals of the present study?

The **Materials and Methods** section describes, in full sentences and well-developed paragraphs, **how the study or experiment was done**. The author provides sufficient detail to allow another appropriately trained scientist to repeat the experiment. Volume, mass, concentration, growth conditions, temperature, pH, type of microscopy, statistical analyses, and sampling techniques are critical pieces of information that must be included. When and where the work was carried out is important if the study was done in the field (in nature), but is not included if the study was done in a laboratory. Conventional labware and laboratory techniques that are common knowledge (familiar to the audience) are not explained. In some instances, it is appropriate to cite references instead of describing all the details.

The **Results** section is **where the findings of the experiment are summarized**, without giving any explanations as to their significance (the "whys" are reserved for the Discussion section). A good Results section has two components:

- A *text*, in which the author describes the results in words
- Some form of *visual* that shows the data and supports the text

In the **Discussion** section, the authors **interpret the results** and give possible explanations. The author may:

- Summarize the results in a way that supports the conclusions.
- Compare the results to other studies.
- Describe limitations, possible sources of error, or inconsistencies in the data; this is preferable to concealing an anomalous result.
- Discuss how the work fits into the big picture.
- Describe future extensions of the current work.

In the **Acknowledgments** section of published research articles, the authors recognize technicians, colleagues, and others who have contributed to the research or production of the paper. In addition, the authors acknowledge the organization(s) that provided funding for the work as well as individuals who provided non-commercially available products or organisms.

References list the **outside sources** the authors consulted in preparing the paper. No one has time to return to a state of zero knowledge and re-discover known mechanisms and relationships. That is why scientists rely so heavily on information published by their colleagues. References are typically cited in the Introduction and Discussion sections of a scientific paper, and the procedures given in the Materials and Methods section are often modifications of those in previous work.

3.3 How to Read Journal Articles

The IMRD structure of journal articles facilitates finding information. However, rather than reading journal articles from start to finish, scientists typically preread to determine initial interest and then read sections selectively (University of Minnesota 2014; Lockman 2012). Specifically, scientists will look for the objectives and conclusions of the paper first. Once they have an overview of the paper, they will check how the results support the conclusions and what methods were used to collect the results. Depending on why scientists were reading the article, they may spend more time on some sections than others.

Papers in scientific journals are written by experts in the field. Because you are not yet an expert, you will probably find it difficult to read and understand journal articles. Even experts read journal articles several times before they understand the methodology and the results. In the following sections, we will apply best practices of active reading to reading journal articles. These practices include asking questions and seeking answers, taking notes, and explaining the main ideas to your peers or your professor (see references under "Academic Skills" in the Bibliography). Ultimately, the goal is to be able to understand the main ideas and express them in your own words.

Prereading

Prereading is also called *previewing* or *surveying*. This method involves skimming a text to get an overview of the content and to determine if the specific information you seek is present. Prereading is designed for speed, not comprehension. Start by skimming the title, the abstract, the key words (if present), the first few sentences of the introduction, and the authors' conclusions in the discussion. If the paper seems promising, look for information related to your research question by section.

Acquire background information on the topic

It's quite possible that you will not understand everything you read. For convenience, you may start looking for background information on a topic by entering key words in Google, Wikipedia, or even YouTube. However, as explained in Section 2.3, these websites should not be considered authoritative sources for academic work. A better choice may be your textbook, written by scientists and reviewed by other scientists before publication. Because textbook authors generally write for a student audience, not a group of experts, your textbook will be easier to read than the primary literature. See Section 3.5 for ways to read biology textbooks efficiently.

Formulate questions

Active reading means reading with a purpose. Scientists read journal articles specifically to acquire the most up-to-date knowledge about a topic directly from the researchers who did the work. In other words, like those scientists, you are reading a journal article to find answers to specific research questions. The following questions, divided according to section, will help you read journal articles with focus.

For the introduction   The structure of the introduction is broad to specific. The first few sentences are aimed at attracting reader interest, and the topic is introduced in general terms. Subsequent sentences narrow down the topic, setting the stage for the specific goals of the study, which are usually stated in the last few sentences.

- What is the general topic of this paper?
- What aspect of this topic is being studied?
- What was already known about the specific topic?
- What was unknown or what questions were the authors trying to answer?
- What was the authors' approach?
- Did the authors propose any hypotheses?
- What specialist terminology (jargon) do I need to define?

For the materials and methods

- What was the *general* approach?
- What *specific* methods were used?
- Am I familiar with these methods? (If not, acquire background information from secondary sources.)

For the results   Look at each figure and read the figure caption to determine what kind of results were collected. Results can be descriptive or numerical.

Photographs, gel images, phylogenetic trees, maps, and flow charts typically show descriptive data. Tables may also contain descriptive data. Possible questions to ask about these kinds of visual aids include:

- What is the subject of the figure (or table)?
- Does the figure show a sequence of events? If so, what is that sequence?
- Are there any labeled organelles, structures, or marks? Why are they important?
- Does the picture show a relationship between form and function? If so, what is the relationship?
- Are there any noteworthy patterns? If so, what is the pattern?

Graphs always show quantitative (numerical) data. Look at each graph and identify the variables. By convention, the independent variable (the one the investigator manipulated) is plotted on the x-axis, and the dependent variable (the one that changes in response to the independent variable) is plotted on the y-axis. On bar graphs, one of the variables is typically categorical rather than quantitative.

- What was the relationship between the independent and dependent variables?
- If a hypothesis was tested, was there a difference between the controls and the treatment groups? If so, how were they different?
- Do the graphs include error bars? Were statistical tests used to analyze the data? If so, did the researchers determine that the differences between treatment groups were statistically significant or could the differences have been due to chance?

For the discussion   The structure of the Discussion section is a triangle, narrow at the top and wide at the base. Information flows from specific to broad (just the opposite of the introduction). The first few paragraphs present the results along with the authors' interpretation. In the next part of the discussion, the results are compared with those in other research papers. The discussion often wraps up with the authors' main conclusions or how this work contributes to the body of knowledge on this topic.

- What were the main results?
- What do the results mean?
- What was the authors' most important conclusion?

Read selectively

With your list of questions in hand, you are now ready to read the article with a specific goal: to find the answers to your questions. Tech-savvy readers who have downloaded PDF versions of journal articles may open them in Adobe Reader and enter key words from their questions to search the document for answers quickly and systematically. The electronic highlighter can be used to mark the phrases that contain information relevant to your research question. In addition, the comments and sticky notes tools are useful for writing notes that help you organize the information and identify questions.

When reading a hard copy, underline the main ideas, and write notes in the margins in your own words. Like the electronic annotation tools, the purpose of marking up a journal article is to help you find information, point out phrases that need clarification, and organize the topics for your research question.

Make sense of the information

As explained in Section 2.3, reading for comprehension involves repeatedly comparing and contrasting new information with prior knowledge. How does this new information fit in, not only with what you already know, but with information that you read in other sources? Are there any apparent contradictions that require closer reading?

If you can explain it in your own words, you understand it. The best way to check your understanding is, first, have a conversation with yourself and listen to what you're saying. Does it make sense? Second, explain your ideas to a classmate or your professor. Doing so quickly identifies gaps in your knowledge and clears up points of confusion.

Share your knowledge

It's no secret that sharing what you've learned with others reinforces your own knowledge. That said, you probably had a specific reason for reading a journal article, and that reason determines *how* you share your knowledge. Will you be giving an oral presentation or writing a paper? Who is your audience: your classmates, your professor, a committee that decides whether to fund your research project, or a community group? The knowledge and interests of your audience will determine the level of technical detail you need to provide. What is the purpose of your communication: to inform, persuade, or entertain your audience? All of these considerations will affect how you share your knowledge.

In the next section, we will learn about two different genres that are used to communicate scientific information: *scientific writing* and *science writing*.

3.4 How Scientific Writing Differs from Science Writing

Scientists typically *read* journal articles to

- Stay current in their field
- Build on what is already known
- Replicate an experiment to verify the results
- Compare their results with those of other researchers
- Improve a method or adapt a method to a different research question
- Improve a product

When scientists *write* journal articles, their audience is other scientists, often experts in the field. When scientists write research proposals, they must convince their audience, not all of whom are experts, that their research is worth funding. Journal articles, research proposals, scientific posters and oral presentations at conferences, and lab reports all fall into the *scientific writing* genre. While each type of writing must be tailored to a specific audience, scientific writing tends to be formal, objective, factual, and structured.

On the other hand, *science writing* is intended for a general audience. This audience is more interested in how science is relevant to their lives than specific experimental details. For the most part, readers of science news articles enjoy learning about discoveries that could hurt or benefit human beings, the wonders of nature, or the amazing accomplishments of scientists (Fahnestock 1986). To engage a general audience, writers must weave scientific facts into an interesting story.

White-nose syndrome is an infectious disease that has killed millions of hibernating bats across North America since it was first observed in a cave near Albany, New York in 2006. We will compare a journal article (JA) by Reeder et al. (2012)* with a science news (SN) article in *Smithsonian* (Nijhuis 2011)** to highlight important differences between the scientific writing and science writing genres. **Each of the headings below describes an**

---

* D. M. Reeder et al. 2012. Frequent Arousal from Hibernation Linked to Severity of Infection and Mortality in Bats with White-Nose Syndrome. *PLOS* 7(6): e38920.

** M. Nijhuis. 2011. Crisis in the Caves. *Smithsonian* 42(4): 66–74.

attribute of science writing. For instructions on how to write lab reports, a form of scientific writing, please see Chapter 5.

Make the title and lead memorable

In journal articles, the title needs to be accurate and descriptive to allow scientists to determine whether the paper is likely to contain the information they seek. On the other hand, science news headlines and **leads** (the first few sentences of an article) need to capture the reader's attention. Entertaining or dramatic titles or ones that pique your curiosity are good choices.

> JA: Frequent Arousal from Hibernation Linked to Severity of Infection and Mortality in Bats with White-Nose Syndrome

> SN: Crisis in the caves. Can scientists stop a new disease that is killing bats in catastrophic numbers?

Emphasize the significance to human beings instead of methods and results

Journal articles are divided into sections, which are arranged in the same order as the elements of the scientific method. New knowledge is built on existing knowledge, and so the Introduction section provides context for the current study; the methods are described to show how the results were obtained; the results lead to certain conclusions. Because validating the results plays such an important role in science, the methods and results sections make up a large portion of journal articles. Indeed, the methods and results sections comprise 43% of the journal article by Reeder et al. In contrast, only 13% of the science news article touches on methods and results, and then only in a general way.

So what content is getting a lot of attention? The scientists (21%), the caves and the surrounding countryside (10%), and why humans should be concerned about a disease that infects bats (9%). We learn that two of the scientists have bat-themed tattoos; we learn how experiences in their youth led them to their current research on bats. And a few well-chosen quotes show that the scientists are human and that they really care about their work.

> SN: "Look at you, with your dirty, dusty little face," Barton coos,…"
>
> "On my worst days, I feel like we're working our tails off just to document an extinction," says Reeder. "But somehow in really teasing apart all of this, in really understanding how they die and why, we may find something

> really important, something we didn't predict,
> something that might help."

For general audiences to want to keep reading a science news article, the importance of the problem or discovery has to be stated explicitly. In contrast, specialists already understand the significance; they are more interested in new findings and how they were obtained. Accordingly, significance is given barely a passing mention in the discussion section of the journal article.

> JA: Insectivorous bats perform significant eco-
> system services because they are primary
> predators of nocturnal insects [35–37]. As
> such, we believe that the loss of cave dwelling
> hibernating bats in North America will be
> ecologically significant.

> SN: But bat biologists say the consequences of
> the North American die-off stretch far be-
> yond the animals themselves. For instance,
> one million bats—the number already felled
> by white-nose syndrome—consume some
> 700 tons of insects, many of them pests,
> every year. Fewer bats mean more mos-
> quitoes, aphids and crop failures. A study
> published in Science this spring estimated
> that bats provide more than $3.7 billion in
> pest-control services to U.S. agriculture
> every year.

Tell a story

Journal articles are divided into sections to allow scientists to find information quickly. However, science news articles are usually not constrained by sections, because the general audience is mainly reading for pleasure. While readers expect the journalistic questions *what, who, when, where, why,* and *how* to be answered in due course, they also want to be entertained. The challenge for the science writer is, therefore, to weave the journalistic questions into a story.

Make the story readable

Science news articles are more readable than journal articles. We can identify specific techniques that science writers use to increase readability (AGU Essential Tips and Tools 2020):

- A conversational tone that informs without lecturing
- Present tense to impart a sense of action and immediacy
- Active voice instead of passive voice
- Nontechnical language instead of jargon and acronyms
- Examples that make numbers relatable
- Analogies that provide context
- Vivid descriptions that evoke an emotional response
- Quotations that demonstrate opinion, enthusiasm, humor, and a can-do attitude

Each of these techniques will be described below.

Use a conversational tone

Some scientists consider science journalism to be "dumbing down" science instead of making science more accessible to a wider audience. While it is true that the science has to be simplified to make it understandable to nonspecialists, it is also important to remember the big picture: an informed public is more likely to make smart decisions. For that reason, using a conversational tone, eliminating unnecessary technical details, and empowering readers with usable knowledge are key elements of science writing for a general audience.

Use present tense

In research articles, present and past tense have specific connotations. Present tense is used for statements that have been accepted as fact by the scientific community. Past tense is used when the statement refers to a specific situation, such as the results obtained in the current study.

> **PRESENT:** Bats with WNS display [present tense] a number of aberrant behaviors… (Reeder et al. 2012 Introduction section)

> **PAST:** During the course of this study, when bats aroused from torpor, they remained [past tense] at euthermic temperatures for a short period, averaging 78.3 ± 27.3 min. (Reeder et al. 2012 Results section)

In the first example, many researchers have documented aberrant behaviors in bats with white-nose syndrome, so present tense is used to indicate a scientific fact. In the second example, the authors use past tense to indicate that this was the result they got in the current study; they do not presume that all studies will get the same result.

Use active voice

In active voice, the subject performs the action, whereas in passive voice, the subject receives the action. Active voice makes sentences more dynamic and often shorter, but passive voice is advantageous in some instances. For example, in the methods section of a journal article, it is implied that the authors carried out the procedure. Passive voice places the emphasis on *what was done*, not who did it.

ACTIVE: We programmed temperature-sensitive dataloggers.

We collected wing skin samples.

We recovered usable data for our analyses.

PASSIVE: Temperature-sensitive dataloggers were programmed...

Wing skin samples (approximately 3 mm × 3 mm each) were collected...

Usable data for our analyses were recovered...

(Reeder et al. 2012 Materials and Methods section)

Eliminate jargon and acronyms

Scientists use **jargon**—specialized, technical vocabulary—and acronyms legitimately to describe complex ideas concisely and efficiently to other specialists. However, jargon is a buzzkill for a general audience. Not only is jargon hard for readers to process, it makes them feel like outsiders, even when the expression is defined in the article. Bullock et al. (2019) found that readers who have difficulty processing the information also feel less confident about their knowledge of science, which makes them less likely to engage in conversations about science. Apparently, the author of the *Smithsonian* article was well aware of the negative consequences of using jargon, because she wrote out *white-nose syndrome* every time.

TECHNICAL TERMS IN JA: White-nose syndrome (WNS), hibernacula, insectivorous bats, cutaneous infection, *Geomyces destructans* (Gd), body temperature (T_b), body mass index (BMI), euthermic T_b, polymerase chain reaction (PCR), skin temperature (T_{sk})

ALTERNATIVES IN SN: white-nose syndrome, disease, bat-killing fungus, caves, bats consume some 700 tons of insects, body fat, hibernate for the winter, rouse from their winter torpor

Put numbers into context and use audience-relevant analogies

Making readers feel comfortable and confident is also the idea behind making numbers relatable and providing analogies that are relevant to a general audience. Here are some examples of how the *Smithsonian* article provides context for the original information in the journal article.

> JA: White-nose syndrome (WNS) is estimated to be responsible for the deaths of at least 5.7 to 6.7 million hibernating bats in the eastern United States and Canada [1,2].

> SN: Can scientists stop a new disease that is killing bats in catastrophic numbers?
>
> ...one million bats...consume some 700 tons of insects, many of them pests, every year.
>
> ...bats provide more than $3.7 billion in pest-control services to U.S. agriculture every year.

> JA: It is hypothesized that infection by Gd disrupts normal physiological functions, such as water balance [8] or other aspects of hibernation physiology, including use of torpor [9].

> SN: In what's been dubbed the "itch and scratch" hypothesis, some scientists posit that the bats are disturbed by the fungus, which accumulates on their muzzle and wings.

Evoke emotions to engage readers

What gives scientific writing credibility is that the methods, results, and conclusions are described transparently and objectively. By intent, scientists keep their own emotions and opinions out of their scientific publications. However, it is precisely those emotions and opinions that a general audience is interested in. The language in the journal article is factual and objective. By contrast, the author of the *Smithsonian* article (Nijhuis 2011) uses colorful language.

> SN: Little brown bats are hanging onto the rocks, alone or in twos and threes, their fur glistening with moisture. Here and there, a dead bat lies on the ground, the bodies hardly more substantial than dried leaves.

> SN: …dead bats piled up in caves throughout the Northeast. The scientists would emerge filthy and saddened, with bat bones—each as thin and flexible as a pine needle—wedged into their boot treads.

We get a vivid impression of what the disease does to the bats, and can't help but to be horrified.

Use direct quotations from authoritative sources

Direct quotations are almost never used in journal articles. Instead, scientists paraphrase the source text and provide an inconspicuous in-text reference (see Section 4.3). In the examples above, the numbers in brackets refer to sources listed in the References section at the end of the journal article.

On the other hand, direct quotations are a key ingredient in science news articles. A general audience *is* interested in what a scientist has to say and how they say it. For example, in only two years, white-nose syndrome had "spread to 19 states and 4 Canadian provinces." A quotation by an authoritative source, a wildlife biologist for the New York State Department of Environmental Conservation, drives home the speed of the spread: "When it first hit, I thought, 'OK, is there anything we can do to keep it within this cave?'" remembers Hicks. "The next year it was, 'Is there anything we can do to secure our largest colonies?' And then the next year it was, 'Can we keep any of these colonies going?' Now we're asking if we can keep these species going."

Closing remarks

The ability to communicate with different audiences is an important life skill. A general audience is interested in acquiring knowledge, but the presentation has to be framed as an engaging story. The techniques used by science writers to make stories readable for the most part do not apply to journal articles written by scientists. Scientists are primarily interested in methods and results that will validate or refute existing knowledge. To make science credible, it must be presented objectively, with precision, and with caution. These characteristics make journal articles hard for non-specialists to process, but learning how to read and write journal articles is an essential skill required in any STEM career.

3.5 How to Read Your Textbook

In the previous section, we learned how scientific writing differs from science writing. Your textbook is likely to contain elements of both genres, because you belong to multiple audiences. Like a general audience, you enjoy reading about the wonders of science and how science can improve your life. Like scientists, you are curious about how things work and why. Although reading your biology textbook will be more difficult than reading a science news article, it will be less difficult than reading a journal article.

If a textbook is required for your course, you can be sure that your instructor expects you to read it. Keep up with your readings. Before each class, preread the text as described in "Survey the content" below. Go to class, take notes, and then actually *read* the assigned text, focusing on the topics emphasized in class. Reading for comprehension requires your full concentration. Turn off your cell phone and eliminate all other distractions for 30-40 minutes. Then take a short break. Repeat the process. You will find that you can accomplish much more in less time when you focus on one thing at a time.

The following steps follow the **SQ3R method** (survey, question, read, recite, review), which was developed to help students learn and remember what they read (see references under "Academic Skills" in the Bibliography). The process takes more time in the beginning, but it saves you time studying in the long run. The following steps work best with a chapter no longer than 25–30 pages.

Survey the content

This first step in the SQ3R method should not take a lot of time and should be done before class. Look over the assigned pages to get an overview of the content. Figure out the main topics by skimming the chapter title, the introduction, the end-of-chapter summary, and the check-your-understanding questions and problems. Then assess the level of detail by skimming the headings and subheadings as well as the pictures. Finally, scan the text for boldfaced terms, which are often vocabulary words that you are expected to know. Prereading the chapter before class allows you to spend less time writing and more time listening, because you already know what information is covered in your textbook.

Go to class and take notes

Some instructors provide PowerPoint slide decks for their lectures, or your textbook may come with a printed lecture notebook or a link to the website where you can download the figures. Bring these printouts to class to use when taking notes. Write down anything the instructor writes on the board. Write down anything associated with the words "This is important." Develop your own shorthand system for taking notes. If you missed something, insert a big question mark so that you can fill in the missing information later. Your notes form a framework for organizing information about the topic, and they help you identify the key concepts that were emphasized in lecture. With your notes as a reference, reading becomes an exercise in elaborating on details and making connections.

Formulate questions

Before you start reading, ask yourself, "What do I already know about this topic?" Make a list of key concepts that you remember. Keep this list handy so that you can correct any misconceptions after you finish reading.

Now review your notes and formulate questions about the topics. Reading engages the eyes, but thinking about the topics and writing down questions provides your brain with additional sensory input. The more senses you engage in learning, the better your memory recall. Here are some possible questions that will help you engage actively with the material.

- Why is [this topic] important?
- How is [topic 1] related to [topic 2]?
- How is [topic 1] different from [topic 2]?
- What experimental evidence led to our current understanding of this topic?
- Why was this approach to the problem taken? Would another also have worked?

Read selectively

Find the sections in your assigned reading that cover the topics emphasized in class. Fill in the gaps in your notes. Find the answers to your questions. As you do so, note any new questions that arise. Note any points of confusion. Define every word so that you become comfortable with the vocabulary. When symbols and formulas are involved, state in words what the terms mean. Interpret any graphs or other experimental data. Make diagrams and concept maps to help you see how topics are related.

Recite

After you're satisfied that you've found answers to your questions, say them out loud using your own words. The combination of speaking and listening engages two additional senses, enhancing your ability to process and remember the information. The acts of speaking and listening may also help you catch errors of logic and missed connections, especially when done in the presence of your study group or instructor. Repeat and refine your answers until you can recite them with confidence.

Review

Set aside a block of time at regular intervals to review your class notes and your reading notes. If you have trouble remembering all of the information, schedule your personal review sessions at more frequent intervals. Definitely try to answer any "Test Your Understanding" questions that come with your textbook. Work the problems without looking at the answers. When you get a wrong answer, try to pinpoint exactly where you went wrong. Articulating what you don't understand will help your instructor give you the kinds of cues that will allow you to figure out the answer for yourself.

Concept (mind) mapping

A **concept map** (also called a **mind map**) is a type of flow chart that links smaller concepts to a main concept. Mind maps are a way for readers to organize knowledge about a topic visually. Mind mapping is based on the premise that new knowledge must be integrated with existing knowledge before further learning is possible. Without this integration, new knowledge is quickly forgotten and misconceptions in existing knowledge will continue to persist (Novak and Cañas 2008). When used with an active reading strategy such as the SQ3R method described above, mind mapping is a powerful way to think about concepts more deeply and retain the information longer (University of Victoria Counselling Services, Reading and Concept Mapping Learning Module, 2020).

The strategy described here works best with a chapter or section of text no longer than 25–30 pages (Palmer-Stone 2010).

1. Take no more than 25 minutes to:
   - Read the chapter title, introduction, and summary (at the end of the chapter, if present).
   - Read the headings and subheadings.
   - Read the chapter title, introduction, summary, headings, and subheadings again.
   - Skim the topic sentence of each paragraph (usually the first or second sentence).
   - Skim italicized or boldfaced words.
2. Close your textbook. Take a full 30 minutes to:
   - Write down everything you can remember about what you read in the chapter (make a mind map). Each time you come to a dead end, use memory techniques such as associating ideas from your reading to lecture notes or other life experiences; visualizing pages, pictures, or graphs; staring out the window to daydream; and letting your mind go blank.
   - Figure out how all this material is related. Organize it according to what makes sense in your mind, not necessarily according to how it is organized in the textbook. Write down questions and possible contradictions to check on later.
3. Open your textbook. Fill in the blanks in your mind map with a different colored pencil.

4. Read the chapter again, this time normally. Make another mind map.

5. Review the material at regular intervals. If you can't construct your mind map in sufficient detail (i.e., you have forgotten much of what you read), then review more frequently.

3.6 Taking Notes

Taking notes—whether you're reading a paper or listening to a lecture—is a good way to stay focused and engage actively with the material. Reviewing your notes helps you identify the topics that your instructor emphasized, so that you have a better idea of what to study for the exams. As mentioned in Section 3.3, notes on your reading form an outline of the main points, helping you organize your ideas when you share your knowledge with others.

Some of the most common note-taking methods are (Stanford University 2020):

- Outlining. This method works best for well-organized written or oral communications. Each main topic is divided into subtopics that relate to and support the main topic.
- Cornell. This method works well for taking notes in class. Draw a horizontal line across the page about 2 inches from the bottom. Draw a vertical line down the page about 2½ inches from the left edge. The top right area is where you take notes. The top left area is for reducing your notes to key concepts (after class). The bottom section is for summarizing the key concepts on each page of notes (after class).
- Annotation. For this method, you jot down notes in the margins of your textbook, journal article, or lecture handouts. This method was applied to journal articles in Section 3.3 "Read Selectively."
- Concept (mind) mapping. This method is useful for visualizing relationships between topics. This method was described in the preceding section.

Here are some practical suggestions for taking notes from Hofmann (2019), Lannon and Gurak (2020), McMillan (2021), Pechenik (2016), and other authorities on scientific writing:

- Don't take notes until you have read the source text at least twice.
- Don't look at the source text when you are taking notes.
- Use your own words and write in your own style.
- Don't use full sentences.

- Develop your own abbreviations (for example, b/c = because and w/o = without) and don't worry about spelling and grammar.
- Write down the most important ideas and how they are related. Look for signal phrases such as "this is important," "this is similar to," "this is different from," and "three steps are involved," because they alert you to important transitions.
- When you review your notes, distinguish your own ideas and questions from those of the original text (e.g., "Me: Applies only to prokaryotes?").
- Use quotation marks to indicate exact or similar wording. Keep in mind that you will have to put the information into your own words if you use the information in your paper.
- Don't cite out of context. Preserve the author's original meaning.
- Give yourself permission to not understand everything. If it's important, get help.
- Fully document the source for later listing in the end references.

By following these best practices for taking notes, your comprehension improves and you reduce the risk of accidental plagiarism (see Section 4.2).

Check Your Understanding

1. Download a journal article that interests you or is relevant to your class. Write down the sections in order and jot down some distinguishing features of each section.

2. Take notes on the journal article. Use your notes to write a summary of the journal article for your classmates.

3. Take notes on the journal article. Use your notes to write a summary of the journal article for a general audience.

4. Take notes on a paragraph, page, or section in your biology textbook. Explain the main ideas to a classmate.

5. Take notes on a paragraph, page, or section in your biology textbook. Explain the main ideas to a friend, who is not a scientist.

6. Compare a science news article with a journal article on the same topic. Make a list of features that characterize each genre. Contrast the two forms of science communication in terms of these features.

7. Write a science news article for a general audience based on a journal article.

Go to **macmillanlearning.com/knisely6e**
and select "Student Site" to access
samples, template files, and tutorial videos

## Chapter 4

Documenting Sources

Objectives

4.1 Distinguish between what kind of information needs to be cited and what does not

4.2 Learn how to paraphrase text to avoid plagiarism

4.3 Distinguish between in-text reference styles for scientific writing and science writing

4.4 Distinguish between in-text references in the Council of Science Editors' Name-Year, Citation-Sequence, and Citation-Name systems

4.5 Learn how to format end references for different kinds of published and unpublished sources in the Name-Year, Citation-Sequence, and Citation-Name systems

Documenting (or citing or acknowledging) your sources is an important part of academic writing. By citing other people's work, you give them credit for ideas that are not your own. At the same time, you give your own work credibility by substantiating your claims and giving readers the opportunity to access references that provide more information about a particular topic.

4.1 Information that Needs to be Cited

General information that is obtained from sources such as news media, textbooks, and encyclopedias does **not** have to be cited.

> EXAMPLE: Most of the ATP in eukaryotic cells is produced in the mitochondria.

Information **that is common knowledge for your audience** does **not** have to be cited. In an introductory course in cell and molecular biology, for example, students would be expected to know that ATP synthase is the enzyme that produces ATP through oxidative phosphorylation.

> EXAMPLE: ATP is synthesized when protons flow down their electrochemical gradient through a channel in ATP synthase.

Information that falls into any of the following categories **must** be cited:

- Information that is not widely known
- Controversial statements, opinions, or other people's conclusions
- Pictures or illustrations that you use but did not produce
- Statistics or formulas used in someone else's work
- Direct quotations

In these situations, if the exact wording is used, it must be enclosed in quotation marks and the source must be cited. Direct quotations are often used in scholarly papers in the humanities, but almost never in scientific writing. This idiosyncrasy of scientific writing requires you to paraphrase the information in the source document. Even when the main ideas are paraphrased, it is still necessary to cite the source (see Section 4.2). Not citing the source constitutes **plagiarism**. Plagiarism is ethically wrong and demonstrates a lack of respect for members of your academic community (faculty and fellow students) and the scientific community in general. Many instructors are now using plagiarism checking services such as Turnitin® and SafeAssign™ by Blackboard to discourage *intentional* plagiarism, such as "borrowing" portions of another student's work, recycling lab reports from previous years, and buying papers on the internet. Plagiarists who are caught can expect to receive at a minimum a failing grade on the assignment and close scrutiny in subsequent work. Plagiarism may also be cause for expulsion from school.

Many cases of plagiarism are *unintentional*, however, and stem from not understanding what is meant by acceptable paraphrasing. The following section provides some guidance.

4.2 Paraphrasing

Paraphrasing means using your own words to express someone else's ideas. Before you can paraphrase, you have to understand the meaning of the source text. If you don't, then you'll just end up copying, which is unintentional plagiarism. Three examples of plagiarism are shown in Table 4.1. In the first example, the writer did not bother to paraphrase at all; she simply copied the source text. In the second example, the writer copied large chunks of the source text, keeping the same basic sentence structure. Even if the words were replaced with synonyms, it would still be considered plagiarism if the sentence structure were not also changed. The third example is an acceptable restatement of the original, but the writer neglected to cite the source. Here are some suggestions for paraphrasing (Indiana University 2020, Purdue OWL 2020):

TABLE 4.1 Examples of plagiarism

Original Text

F_1 extends from the membrane, with the α and β subunits alternating around a central subunit γ. ATP synthesis occurs alternately in different β subunits, the cooperative tight binding of ADP + P_i at one catalytic site being coupled to ATP release at a second. The differences in binding affinities appear to be caused by rotation of the γ subunit in the center of the $\alpha3$ $\beta3$ hexamer.

Plagiarized text	Reason
According to Fillingame (1997), F_1 extends from the membrane, with the α and β subunits alternating around a central subunit γ. ATP synthesis occurs alternately in different β subunits, the cooperative tight binding of ADP + P_i at one catalytic site being coupled to ATP release at a second. The differences in binding affinities appear to be caused by rotation of the γ subunit in the center of the $\alpha3$ $\beta3$ hexamer.	The author's actual words were used without quotation marks or indenting the citation. Because direct quotations are not used in scientific papers, it is imperative that you paraphrase. Using the original text is plagiarism even when the source is cited.
F_1 consists of α and β subunits alternating around a central subunit γ. In the β subunits, tight binding of ADP + P_i occurs at one catalytic site and ATP is released at a second. The different binding affinities may be caused by rotation of the γ subunit in the center (Fillingame 1997).	The basic sentence structure of the original text was maintained. A few words were omitted or changed, but the text is still highly similar to the original.
ATP synthase consists of a transmembrane protein (F_o), a central shaft (γ), and an F_1 head made up of α and β subunits. As protons enter F_o, the shaft rotates, changing the conformation of the β subunits, allowing ADP and P_i to bind and be released as ATP.	The text was paraphrased, but the source of the information was not cited.

Source: R. H. Fillingame. 1997. *J. Exp. Biol.* 200: 217–224, accessed 2017 Jan 19.

- Read the text several times to make sure you understand the meaning.
- Determine how you are going to use the information to make your point.
- Do not look at the source text when you restate the main ideas in your own words.
- Say your ideas out loud or, better yet, explain them to a classmate. If you can explain them, then you understand them.
- Compare your paraphrased text to the original. If they are too similar, revise your paraphrase.
- Cite the source, as described in Section 4.3.

4.3 Council of Science Editors (CSE) Reference Styles

The Council of Science Editors (CSE) is a professional society that provides guidelines for publishing in the sciences. The CSE Scientific Style and Format Manual is to scientific publishing what the MLA Handbook is to the humanities and the APA Manual is to the social sciences. There are many other style guides, so it's a good idea to confirm your instructor's requirements. If you are preparing a manuscript for publication in a journal, be sure to follow the journal's instructions for authors.

The CSE Manual (2014) describes three systems for documenting sources: Citation-Sequence System (C-S), Citation-Name System (C-N), and Name-Year System (N-Y).

> **Citation-Sequence System**. In the text, the source of the cited information is provided in an abbreviated form as a superscripted endnote or a number in square brackets or parentheses. On the references pages that follow the Discussion section, the sources are listed in **numerical order** and include the full reference.

> **Name-Year System**. In the text, the source is given in the form of author(s) and year. On the references pages that follow the Discussion section, the references are listed in **alphabetical order** according to the first author's last name.

> **Citation-Name System**. This system is a hybrid of the Citation-Sequence and Name-Year systems. In the text, the source of the cited information is provided in an abbreviated form as a superscripted endnote or a number in square brackets or parentheses. On the references pages that follow the Discussion section, the references are listed in **alphabetical order** according to the first author's last name. The references are then numbered sequentially.

In all three systems, the *in-text reference* is intended to be inconspicuous. A superscripted number or a number in parentheses (C-S and C-N systems) or authors' names and year (N-Y system) are minimally disruptive to the flow of the sentence. There is no need to mention the authors' credentials or affiliations, because the peer review process has already vetted their work.

C-S OR C-N IN-TEXT STYLE:	Fungal hyphae were found under the skin of bats infected with WNS[1]. *or* Fungal hyphae were found under the skin of bats infected with WNS [1].
N-Y IN-TEXT STYLE:	Fungal hyphae were found under the skin of bats infected with WNS (Meteyer et al. 2009).

SCIENCE NEWS IN-TEXT STYLE:	Dr. Carol Meteyer, a scientist at the National Wildlife Health Center, US Geological Survey, in Madison, Wisconsin, explains in a recent article in Journal of Veterinary Diagnostic Investigation that fungal hyphae were found under the skin of bats infected with white-nose syndrome. [**For scientific writing, do not include the author's title and affiliation or the name of the publication in the in-text reference.**]

Another difference between citations in science writing and scientific writing is that direct quotations are almost never used in the latter. Instead, restate the information from the source text in your own words and then cite the source (see Section 4.2).

N-Y IN-TEXT STYLE:	Fungal hyphae were found under the skin of bats infected with WNS (Meteyer et al. 2009).

SCIENCE NEWS IN-TEXT STYLE:	Carol Meteyer, a scientist at the National Wildlife Health Center, explains that "fungal hyphae form cup-like epidermal erosions and ulcers in the wing membrane and pinna with involvement of underlying connective tissue." [**Do not use direct quotations in scientific writing.**]

With regard to the *end reference*, the systems differ in the sequence of information and the listing of the month of publication. In the N-Y system, the year of publication follows the authors' names; in the C-S and C-N systems, the year follows the journal name. The month of publication is only used in the C-S and C-N systems.

The Name-Year system has the advantage that people working in the field will know the literature and, on seeing the authors' names, will understand the in-text reference without having to check the end reference. With the Citation-Sequence and Citation-Name systems, the reader must turn to the reference list at the end of the paper to gain the same information. Regardless of which system you use, learn the proper way to format both the in-text reference and the end reference and use one system consistently throughout any given paper.

The Name-Year system

The *in-text reference* consists of author(s) and year. The author(s) may be cited in parentheses at the end of the sentence or they may be the subject of the sentence, as shown in the following examples:

The Name-Year System

TABLE 4.2 The number of authors determines how the source is cited in N-Y system

Number of Authors	Author as Subject	Parenthetical Reference (The comma between author[s] and year is optional.)
1	Author's last name (year) found that...	(Author's last name, year)
2	First author's last name and second author's last name (year) found that ...	(First author's last name and second author's last name, year)
3 or more	First author's last name followed by "and others" or *et al.* (year) found that ...	(First author's last name and others, year) or *et al.* instead of *and others*

Note: If you cite more than one paper published by the same author in *different* years, list them in chronological order: (Dawson 2001, 2003). If you cite more than one paper published by the same author in the *same* year, add a letter after the year: "...was described in recent work by Dawson (1999a, 1999b)."

PARENTHESES: C-fern gametophytes respond to antheridiogen only for a short time after inoculation (Banks and others 1993).

AS THE SUBJECT: Banks and others (1993) found that C-fern gametophytes respond to antheridiogen only for a short time after inoculation.

The **number of authors** determines how the *in-text reference* is written in the N-Y system (Table 4.2). For *one* author, write the author's last name and year. For *two* authors, write both authors' last names separated by the word *and* followed by the year. For *three or more* authors, write the first author's last name, the words *and others* (or *et al.*), and then the year.

In the *end references*, the sources are listed in **alphabetical order** according to the first author's last name. The format of the source determines which elements are included (Table 4.3). When there are 10 or fewer authors, list all authors' names. When there are more than 10 authors, list the first 10 and then write *et al.* or *and others* after the tenth name. For each reference, list the authors' names in the order they appear on the title page. Write each author's name in the form Last name First initials. Use a comma to separate one author's name from the next. Use a period only after the last author's name.

Examples of in-text references and their corresponding end references are given in Table 4.4. See The CSE Manual (2014) and Patrias (2007-present) for examples of many other kinds of sources.

The Citation-Sequence system

The *in-text reference* consists of a superscripted endnote (never a footnote) or a number in parentheses or square brackets within or at the end of the paraphrased sentence. The first reference cited is number 1, the second reference cited is number 2, and so on.

TABLE 4.3 General format of two systems of source documentation used in scientific papers

Name-Year End Reference System

The references are listed in **alphabetical order**. The last name is written first, followed by the initials. When there are 10 or fewer authors, list all authors' names. When there are more than 10 authors, list the first 10 and then write *et al.* or *and others* after the tenth name. Type references with hanging indent format.

Journal article	First author's last name First initials, Subsequent authors' names separated by commas. Year of publication. Article title. Journal title Volume number(issue number): inclusive pages.
Article in book	First author's last name First initials, Subsequent authors' names separated by commas. Year of publication. Article title. In: Editors' names followed by a comma and the word *editors*. Book title, edition. Place of Publication: Publisher. pp inclusive pages.
Book	First author's or editor's last name First initials, Subsequent authors' or editors' names separated by commas. Year of publication. Title of book. Place of Publication: Publisher. Total number of pages in book followed by *p*.

Citation-Sequence End Reference System

The references are listed **in the order they are cited**. The author's last name is written first, followed by the initials. When there are 10 or fewer authors, list all authors' names. When there are more than 10 authors, list the first 10 and then write *et al.* or *and others* after the tenth name.

Journal article	Number of the citation. First author's last name First initials, Subsequent authors' names separated by commas. Article title. Journal title Year Month; Volume number(issue number): inclusive pages.
Article in book	Number of the citation. First author's last name First initials, Subsequent authors' names separated by commas. Article title. In: Editors' names followed by a comma and the word *editors*. Book title, edition. Place of Publication: Publisher; Year of publication. pp inclusive pages.
Book	Number of the citation. First author's or editor's last name First initials, Subsequent authors' or editors' names separated by commas. Title of book. Place of Publication: Publisher; Year of publication. Total number of pages in book followed by *p*.

SUPERSCRIPTED ENDNOTE:	There are four commonly used methods for determining protein concentration: the biuret method[1], the Lowry method[2], the Coomassie Blue (CB) dye-binding method[3], and the bicinchoninic acid (BCA) assay[4].
PARENTHESES:	The Kjeldahl procedure is time-consuming and requires a large amount of sample (1, 2).
BRACKETS:	Several review articles compare the advantages and disadvantages of these protein assays [5–10].

TABLE 4.4 Examples of in-text citation and end reference format of two systems of source documentation used in scientific papers

Name-Year System

IN-TEXT REFERENCES

3 or more authors	Gametophytes of the tropical fern *Ceratopteris richardii* (C-fern) develop either as males or hermaphrodites. Their fate is determined by the pheromone antheridiogen (Näf 1979; Näf and others 1975). Banks and others (1993) found that gametophytes respond to antheridiogen only for a short time between 3 and 4 days after inoculation. Although the structure of antheridiogen is unknown, it is thought to be related to the
2 authors	gibberellins (Warne and Hickok 1989). Gibberellins are a group of plant hormones involved in stem elongation, seed germination, flowering, and
1 author	fruit development (Treshow 1970).

CORRESPONDING END REFERENCES

Journal article	Banks J, Webb M, Hickok L. 1993. Programming of sexual phenotype in the homosporous fern *Ceratopteris richardii*. Inter. J. Plant Sci. 154(4): 522-534.
Article in book	Näf U. 1979. Antheridiogens and antheridial development. In: Dyer AF, editor. The Experimental Biology of Ferns. London: Academic Press. pp. 436-470.
Journal article	Näf U, Nakanishi K, Endo M. 1975. On the physiology and chemistry of fern antheridiogens. Bot. Rev. 41(3): 315-359.
Book	Treshow M. 1970. Environment and Plant Response. New York: McGraw-Hill. 250 p.
Journal article	Warne T, Hickok L. 1989. Evidence for a gibberellin biosynthetic origin of *Ceratopteris* antheridiogen. Plant Physiol. 89(2): 535-538.

Citation-Sequence System

IN-TEXT REFERENCES

Sources are listed in the order they are cited	Gametophytes of the tropical fern *Ceratopteris richardii* (C-fern) develop either as males or hermaphrodites. Their fate is determined by the pheromone antheridiogen (1, 2). Gametophytes respond to antheridiogen only for a short time between 3 and 4 days after inoculation (3). Although the structure of antheridiogen is unknown, it is thought to be related to the gibberellins (4). Gibberellins are a group of plant hormones involved in stem elongation, seed germination, flowering, and fruit development (5).

CORRESPONDING END REFERENCES

Article in book	1. Näf U. Antheridiogens and antheridial development. In: Dyer AF, editor. The Experimental Biology of Ferns. London: Academic Press; 1979. pp. 436-470.
Journal article	2. Näf U, Nakanishi K, Endo M. On the physiology and chemistry of fern antheridiogens. Bot. Rev. 1975 Jul-Sep; 41(3): 315-359.
Journal article	3. Banks J, Webb M, Hickok L. Programming of sexual phenotype in the homosporous fern *Ceratopteris richardii*. Inter. J. Plant Sci. 1993 Dec; 154(4): 522-534.
Journal article	4. Warne T, Hickok L. Evidence for a gibberellin biosynthetic origin of *Ceratopteris* antheridiogen. Plant Physiol. 1989 Feb; 89(2): 535-538.
Book	5. Treshow M. Environment and Plant Response. New York: McGraw-Hill; 1970. 250 p.

Pagination is optional. If present, this is the total number of pages in the book, not the pages from which information was cited.

In the *end references*, the sources are listed in **numerical order** (in the order of citation). The format of the source determines which elements are included (Table 4.3). When there are 10 or fewer authors, list all authors' names. When there are more than 10 authors, list the first 10 and then write *et al.* or *and others* after the tenth name. Write each author's name in the form Last name First initials. Use a comma to separate one author's name from the next. Use a period only after the last author's name.

Examples of in-text references and their corresponding end references are given in Table 4.4. See The CSE Manual (2014) and Patrias (2007-present) for examples of many other kinds of sources.

The Citation-Name system

In the *end references*, the sources are listed in **alphabetical order** according to the first author's last name. The year and month of publication follow the journal name, as in C-S end reference format. The references are then numbered sequentially so that the first reference is number 1, the second reference is number 2, and so on. The *in-text references* consist of superscripted endnotes (never footnotes) or a number in parentheses or square brackets within or at the end of the paraphrased sentence.

4.4 Online Publications

There is hardly a printed scientific publication that does not also have an online presence. Electronic publication not only allows current research to be disseminated more quickly, it allows timely corrections and updates to be made after publication. Although many of us prefer to read a hard copy, the reality is that we will be citing resources that we access online.

In Section 4.3, you learned that the in-text reference and end reference format differ for journal articles, articles in a book, and books. These differences apply to both print and online publications. For a journal article, therefore, you should be able to locate on the website the names of the authors, a title, the journal name, a date of publication, the volume and issue number, and the extent (number of pages or similar). Besides this basic information, the CSE Manual (2014) recommends that you provide two additional items when your reference comes from the internet: the URL (uniform resource locator) and the date accessed. **For your lab reports, it is sufficient to treat references obtained online as print sources** (unless your instructor tells you otherwise). If you would like to publish your work in a journal that adheres strictly to CSE guidelines, however, the following sections show the reference format for some of the most common online resources. For a comprehensive discussion of internet citation formats along with many examples, see Patrias (2007–present).

Figure 4.1 The same online journal article can be obtained from different pro- ▶ viders, depending on your academic library's subscriptions. Both providers give the information needed for the full reference of the journal article (in boxes): Author(s). Date of publication. Title of article. Title of journal. [date updated; date accessed]; Volume(issue): Inclusive page numbers. URL or DOI. (A) BioOne Complete includes the digital object identifier (DOI), which is a persistent link to the online article. The DOI may be used in addition to the URL. (B) ProQuest has a Cite button that makes it easy to copy the full reference in the desired format. The reference can also be exported to a citation manager such as RefWorks, EndNote, EasyBib, Mendeley, or Zotero.

The problem with sources on the internet is that they may disappear at any time or their URL may change. To provide a persistent link to online articles and books, many publishers include a DOI (digital object identifier) on the first page of the publication (Figure 4.1). The DOI consists of a unique string of numbers and letters that, when pasted into a browser, leads direct- ly to that publication. If available, the DOI may be included in addition to the URL. The goal is to get the reader to the source quickly and reliably.

> When URLs are used in text, they do *not* need special formatting. They do *not* need to be enclosed in angle brackets (< >) and they do *not* need to be underlined and in color. Every character in a URL is significant, as are spaces and capitalization. Very long URLs can be broken before a punctuation mark (tilde ~, hyphen -, underscore _, period ., forward slash /, backslash \, or pipe |). The punctuation mark is then moved to the next line.

The following online publications are listed in alphabetical order by type. The general format for the end reference is given for each type in both name-year (N-Y) and citation-sequence (C-S) style.

Blog

A **blog** is short online post that is updated regularly.

> **N-Y:** Author's name. Year Month Date. Title of post. Title of blog. [accessed date]. URL

> **EX.** Buis A. 2020 Sept 8. Making sense of 'climate sensitivity': New study narrows the range of uncertainty in future climate projections. Ask NASA Climate. [accessed 2020 Sept 28]. https://climate.nasa.gov/blog/

(A)

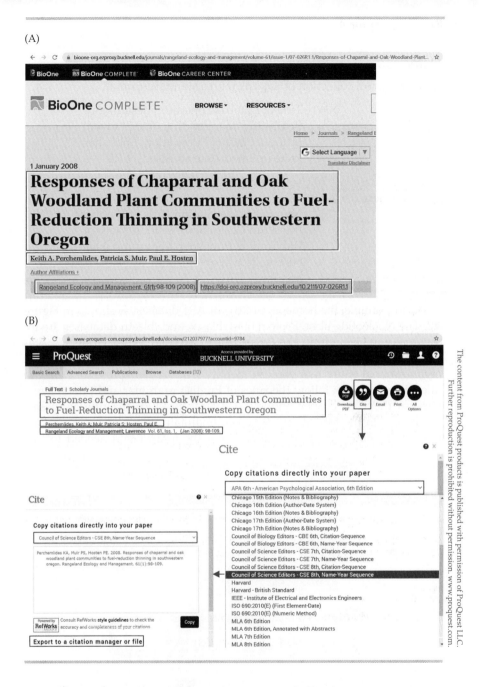

(B)

To cite a blog in the **Citation-Sequence** and **Citation-Name systems**, move the date after the blog title.

C-S: Number of the citation. Author's name. Title of post. Title of blog. Year Month Date. [accessed date]. URL

Database

A **database** is a collection of records with a standard format.

> **N–Y:**   Title of Database [medium designator].
> Beginning date – ending date (if given).
> Edition. Place of Publication: Publisher.
> [date updated; date accessed]. URL

To cite a database in the **Citation-Sequence system**, move the date after the publisher:

> **C–S:**   Number of the citation. Title of Database
> [Medium Designator]. Edition. Place of
> Publication: Publisher. Beginning date –
> ending date (if given). [date updated; date
> accessed]. URL

As an example, the homepage of the BLAST database is shown in Figure 4.2. The Nucleotide blast, Protein blast, blastx, and tblastn databases are individual websites within the larger BLAST website. When citing websites within websites, the following rule applies: Always cite the most specific organizational entity that you can identify (Patrias 2007–present). Database titles do not always follow the rules of English grammar and punctuation. Because they are proper nouns, however, reproduce the title as closely as possible to the format on the screen (maintain upper or lower case letters, run-together words, etc.). Sometimes the information needed for the

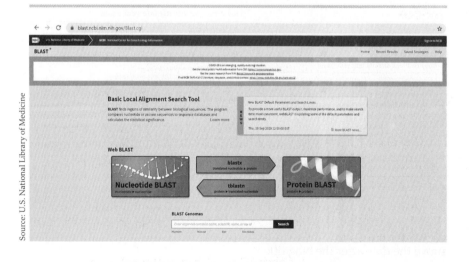

Source: U.S. National Library of Medicine

Figure 4.2   Homepage for the National Center for Biotechnology Information's BLAST database. To search for a specific nucleotide sequence, use Nucleotide BLAST, one of the databases within the BLAST database.

reference may be absent or hard to find. In this example, the beginning to ending dates and the edition of the database are not specified. The location of the place of publication and the publisher can be found at the bottom of the page. Do your best to reference the source with the information provided.

A good faith attempt at citing the Nucleotide blast database in **Name-Year format** would be as follows:

> EX. Nucleotide blast [database on the internet]. Bethesda (MD): U.S. National Library of Medicine, National Center for Biotechnology Information. [accessed 2020 Sept 28]. http://blast.ncbi.nlm.nih.gov/Blast.cgi

The *in-text reference* for a database in **Name-Year format** follows the same principles used for print publications (see Table 4.4) with a minor modification. The author is replaced with the title of the database and, when the date of publication is not known (as in the current example), the order of preference is the copyright date; the date of modification, update, or revision; and the date accessed (CSE Manual 2014). An example of an *in-text reference* in **Name-Year format** for this database would be:

> EX. There was a 100% match between the DNA sequence of Sample 1 and the SV40 sequence in the NCBI databank (Nucleotide blast database [accessed 2020 Sept 28]).

e-Book

> N-Y: Author(s). Date of publication. Title of book. Edition. Place of publication: publisher; [date updated; date accessed]. URL

> EX. Blum D, Knudson M, Henig RM. 2005. A Field Guide for Science Writers: The Official Guide of the National Association of Science Writers. 2nd edition. Oxford University Press USA; [accessed 2020 Sept 28]. https://ebookcentral.proquest.com/lib/bucknell/detail.action?docID=270856

In the **Citation-Sequence** and **Citation-Name systems**, move the date after the publisher.

> C-S: Number of the citation. Author(s). Title of book. Edition. Place of publication: publisher; date of publication [date updated; date accessed]. URL

Image or infographic

To reference an image or infographic in the **Name-Year**, **Citation-Sequence** and **Citation-Name** systems

> REF: Artist's Name. Descriptive Title. City (ST): publisher or producer. [date accessed]. URL

> EX.   Centers for Disease Control and Prevention. 10 things you can do to manage your COVID-19 symptoms at home. Atlanta (GA): CDC. [2020 Sept 28]. https://www.cdc.gov/coronavirus/ 2019-ncov/downloads/10Things.pdf

Journal article

The *in-text reference* for an online journal article is nearly the same as that for a printed journal article (see Table 4.4), with the internet-specific items highlighted below in bold (CSE Manual 2014).

To reference an online journal article in the **Name-Year system**

> N-Y: Author(s). Date of publication. Title of article. Title of journal. **[date updated; date accessed]**; Volume(issue):Inclusive page numbers. **URL and DOI, if available**

A screen shot of an online journal article web page is shown in Figure 4.1, with the elements required for citation in boxes. Do not confuse the title of the journal with the publisher or digital library. BioONE, SAGE, ScienceDirect, and JSTOR are not journal titles. These "providers" do not belong in the reference.

> EX.   Perchemlides KA, Muir PS, Hosten PE. 2008. Responses of chaparral and oak woodland plant communities to fuel-reduction thinning in southwestern Oregon. Rangeland Ecology and Management. [accessed 2020 Sept 28] 61(1):98-109. https://doi-org.ezproxy.bucknell. edu/10.2111/07-026R1.1

In the **Citation-Sequence** and **Citation-Name systems**, move the date after the title of the journal.

> C-S: Number of the citation. Author(s). Title of article. Title of journal. Year and month **[date updated; date accessed]**; Volume(issue): Inclusive page numbers. **URL and DOI, if available**

EX. 1. Perchemlides KA, Muir PS, Hosten PE. Responses of chaparral and oak woodland plant communities to fuel-reduction thinning in southwestern Oregon. Rangeland Ecology and Management 2008 Jan; [accessed 2020 Sept 28] 61(1):98-109. https://doi-org.ezproxy. bucknell.edu/10.2111/07-026R1.1

Newspaper article

N-Y: Author. Date of publication. Article title. Unabbreviated newspaper title [internet]. [date accessed];Section:[about # p. or screens]. URL

EX. Cannon J. 2020 Sept 8. Tahlequah, an orca that carried her dead calf for 17 days and 1,000 miles, gives birth again. USA TODAY [internet]. [accessed 2020 Sept 28];Nation:[about 1 p.] https://www.usatoday. com/story/news/nation/2020/09/08/ tahlequah-orca-southern-resident-killer-whale-gives-birth/5747567002/

In the **Citation-Sequence** and **Citation-Name systems**, move the date after the newspaper title.

C-S: Number of the citation. Author. Article title. Unabbreviated newspaper title [internet]. [date of publication; date accessed];Section:[about # p. or screens]. URL

Podcast or webcast

N-Y: Narrator's Name. Date first aired, length. Title of podcast episode [descriptive word, episode number if available]. Name of podcast show. Producer. [accessed date]. URL

EX. Zimmerman S. 2019 Jan, 5:11 min. How CRISPR lets you edit DNA. TED Talks. TED. [accessed 2020 Sept 28]. https://www.ted.com/ talks/andrea_m_henle_how_crispr_lets_you_ edit_dna

In the **Citation-Sequence** and **Citation-Name systems**, move the date after the producer's name.

C-S: Number of the citation. Narrator's Name. Title of podcast episode [descriptive word, episode number if available]. Name of podcast show. Producer. Date first aired, length. [accessed date]. URL

Social media

N-Y: Username or group/page name. Date and time posted, if available. Social media platform [descriptive word for page type, post type]. [accessed date]. URL

EX.   Chris Martine, PhDuh. 2020 Sept 28. Twitter [Martine Botany Lab] [accessed 2020 Sept 29]. https://twitter.com/MartineBotany

In the **Citation-Sequence** and **Citation-Name systems**, move the date after the name of the platform.

C-S: Number of the citation. Username or group/ page name. Social media platform [descriptive word for page type, post type]. Date and time posted, if available. [accessed date]. URL

Video

N-Y: Title of video [descriptive word, episode number if available]. Date first aired or posted, length. Title of program. Producer. [accessed date]. URL

EX.   The time is ripe to explain the difference between fruits and vegetables. 2015 June 25, 4:08 min. HuffPost Science. Verizon Media. [accessed 2020 Sept 28] https://www.huffpost.com/entry/ difference-fruit-vegetable_n_7664902

In the **Citation-Sequence** and **Citation-Name systems**, move the date after the producer's name.

C-S: Number of the citation. Title of video [descriptive word, episode number if available]. Title of program. Producer. Date first aired or posted, length. [accessed date]. URL

Websites and homepages

A **homepage** is the main page of a website, which provides links to different content areas of the site. Most of the information required to cite a website is found on the homepage. Make sure the organization or individual responsible for the website is reputable and, if possible, confirm information on the site using another source.

> N-Y: Title of Homepage. Date of publication. Edition. Place of publication: publisher; [date updated; date accessed]. URL

To cite a homepage in the **Citation-Sequence** and **Citation-Name systems**, move the date after the publisher.

> C-S: Number of the citation. Title of Homepage. Edition. Place of publication: publisher; date of publication [date updated; date accessed]. URL

An example of a homepage is shown in Figure 4.3. All of the information required to cite this source is readily located. When the date of publication is not specified, the order of preference is the copyright date; the

Figure 4.3 The National Academies' website for information on climate change. Well-constructed websites make it easy to find the title, date, responsible organization, and place of publication.

date of modification, update, or revision; and the date accessed (CSE Manual 2014). In this example, the copyright date, preceded by a lower case *c*, is used in the end reference.

> EX.   Climate Resources at the National Academies.
> c2020. Washington DC: National Academy
> of Sciences; [accessed 2020 Sept 28]. https://
> www.nationalacademies.org/topics/climate

The *in-text reference* for a homepage in **Name-Year format** follows the same principles used for print publications (see Table 4.4) with a minor modification. The author is replaced with the title of the homepage. For the year, the order of preference is the date of publication; the copyright date; the date of modification, update, or revision; and the date accessed (CSE Manual 2014).

4.5 Unpublished Material

Laboratory exercises

If your instructor asks you to cite unpublished laboratory exercises in your laboratory reports, the *end reference* could look like this:

> N-Y: Author (if unknown, replace with title of lab
> exercise). Year. Title of lab exercise. University.
> Department. Course number.

> C-S: Number of the citation. Author (omit if
> unknown). Title of lab exercise. University.
> Department. Course number. Year.

In N-Y format, the *in-text reference* would include the author(s) and year, or, if the author is unknown, the title of the lab exercise and year. The use of *anonymous* is not recommended (CSE Manual 2014).

Lectures, emails, and discussions

Unpublished information that is not common knowledge should be acknowledged when you use it in your scientific communications. The *in-text reference* includes the authority, the date, and the words "personal communication" or "unreferenced." For example:

> Most viruses affecting honey bees have genomes composed of
> RNA rather than DNA (M. Pizzorno, personal communication,
> 2016 Sept 22).

Personal communications are not included in the full references list.

Papers and poster sessions at meetings

N-Y: Author(s). Date of the conference. Title of paper (poster). Paper (poster) session presented at: Title of conference. Name of the conference; place of the conference.

EX. Naughton L, Cannizzaro D, Pask G. 2020 Jan 3-7. One Big, Smelly Family: Decoding the olfactory receptors in the Indian jumping ant. Poster session presented at: SICB 2020. The Society for Integrative & Comparative Biology Annual Meeting; Austin, TX

C-S: Author(s). Title of paper (poster). Paper (poster) session presented at: Title of conference. Name of the conference; date; place.

Preprints and forthcoming articles

Preprints are manuscripts that have not yet been peer reviewed, but have been disseminated to invite comments and establish order of discovery. Many preprints are uploaded to preprint servers such as bioRxiv and ArXiv.

N-Y: Author(s). Date. Title. Repository [Preprint]. [date accessed]. URL and DOI, if available

EX. Shively CA, Appt E, Chen H, Day SM, Frye BM, Shaltout HA, Silverstein-Metzler MG, Snyder-Mackler N, Uberseder B, Vitolins MZ, Register TC. 2020. Mediterranean diet, stress resilience, and aging in nonhuman primates. bioRxiv 2020.09.25.313825 [Preprint]; [accessed 2020 Sept 29] doi: https://doi.org/10.1101/2020.09.25.313825

C-S: Author(s). Title. Repository [Preprint]. Date [date accessed]. URL and DOI, if available

Forthcoming articles are manuscripts that have undergone peer review and have been accepted for publication, but have not yet been published.

N-Y: Author(s). Forthcoming date. Title of article. Title of journal.

EX. Collin R, Dagoberto EV-P, Paulay G, Boyle MJ. 2020 Oct. World travelers: DNA barcoding

> unmasks the origin of cloning asteroid larvae
> from the Caribbean. The Biological Bulletin.
>
> C-S: Author(s). Title of article. Title of journal.
>   Forthcoming date.

Check Your Understanding

1. Paraphrase a paragraph, page, or section of your biology textbook.

2. Paraphrase a paragraph, page, or section of a journal article that interests you.

3. Write a sentence in which you cite information from a journal article using the N-Y system. Then rewrite the sentence for a general audience.

4. Search a recommended database on your library's subject-guide website for a topic of interest (see Chapter 2). Click on titles that seem relevant. Write the end reference for 2-3 journal articles that are relevant to your topic. Choose N-Y, C-S, or C-N format.

5. Search Google or Google Scholar for blogs, images, podcasts, videos, social media, or websites for a topic of interest (see Chapter 2). Click on results that seem relevant. Write the end reference for 2-3 results in N-Y, C-S, or C-N format.

6. Choose an unpublished lab exercise, lecture, or discussion you had with your professor. Summarize the main ideas and cite the source.

Go to **macmillanlearning.com/knisely6e**
and select "**Student Site**" to access
samples, template files, and tutorial videos

## Chapter 5

Step-by-Step Instructions for Preparing a Scientific Paper

Objectives

5.1 Explain why the IMRD sections are not written in the order they appear in the paper

5.2 Describe the purpose, content, and organizational structure of each section

5.3 Recognize common mistakes students may make when writing different sections of lab reports (and avoid them!)

5.4 Explain the significance of tense (past and present) in scientific writing

5.5 Understand how passive voice can shift the emphasis from who carried out the procedure to what was done

5.6 Explain how visuals (figures and tables) are selected based on the type of data

In order to communicate science according to accepted conventions, the following skills are required:

- A solid command of the English language
- An understanding of the scientific method
- An understanding of scientific concepts and terminology
- Advanced word processing skills
- Knowledge of computer graphing software
- The ability to read and comprehend journal articles
- The ability to search the primary literature efficiently
- The ability to evaluate the reliability of internet sources

If you are a first- or second-year college student, it is unlikely that you possess all of these skills when you are asked to write your first laboratory report. Don't worry. The instructions in this chapter will guide you through the steps involved in preparing the first draft of a laboratory report. Revision is addressed in Chapter 7, and the Appendices will help you with word processing and graphing tasks.

5.1 Timetable

Preparing a laboratory report or scientific paper is hard work. It will take much more time than you expect. Writing the first draft is only the first step. You must also allow time for editing and proofreading (revision). If you work on your paper in stages, the final product will be much better than if you try to do everything at the last minute.

The timetable outlined in Table 5.1 breaks the writing process down into stages, based on a one-week time frame. You can adjust the time frame according to your own deadlines.

Format your paper correctly

Although content is important, the appearance of your paper is what makes the first impression on the reader. Before submitting papers electronically, print out and proofread the hard copy. You will be surprised at how formatting and other kinds of errors jump out at you when you read on paper instead of on screen. If your instructor asks you to turn in assignments on paper, make sure the pages are in order and the print is legible. Subconsciously or not, the reader/evaluator is going to associate a sloppy paper with sloppy science. You cannot afford that kind of reputation. In order for your work to be taken seriously, your paper has to have a professional appearance.

Scientific journals specify the format in their "Instructions to Authors" section. If your instructor has not given you specific instructions, the layout specified in Table 5.2 will give your paper a professional look.

Consult the sample "good" student laboratory report in Chapter 8 for an overview of the style and layout. An electronic file called "Biology Lab Report Template," available at **macmillanlearning.com/knisely6e** is formatted according to the guidelines of Table 5.2 and provides prompts that help you get started writing in scientific paper format. For details on how to format documents in Microsoft Word, see Section A1.8 in Appendix 1.

Computer savvy

Know your computer and your word processing software. Most of the tasks you will encounter in writing your paper are described in Appendix 1, "Word Processing in Microsoft Word" and Appendix 2, "Making Graphs in Microsoft Excel." If there is a task that is not covered in these appendices, write it down and ask an expert later. If you run into a major problem that prevents you from using your computer, you should have a backup plan in place (access to another computer).

Always back up your files somewhere other than your computer's hard drive. Options may include a USB flash drive (also called a jump drive

TABLE 5.1	Timetable for writing your laboratory report	
Time Frame	**Activity**	**Rationale**
Day 1	Complete laboratory exercise.	It's fun. Besides, you need data to write about.
Days 2–3	Write first draft of laboratory report.	The lab is still fresh in your mind. You also need time to complete the subsequent tasks before the due date.
Day 4	Revise first draft (hard copy).	Always take a break after writing the first draft and before revising it. This "distance" gives you objectivity to read your paper critically.
Day 5	Give first draft to a peer reviewer for feedback, *if your instructor permits it.*	Your peer reviewer is a sounding board for your writing. He/she will give you feedback on whether what you intended to write actually comes across to the reader. You may wish to alert your peer reviewer to concerns you have about your paper (see "Get Feedback" in Section 7.4).
	Arrange to meet with your peer reviewer after he/she has had time to review your paper ("writing conference").	An informal discussion is useful for providing immediate exchange of ideas and concerns.
Day 6	Peer reviewer reviews laboratory report.	The peer reviewer should review the paper according to two sets of criteria. One is the conventions of scientific writing as described in Section 3.2 and this chapter. The other is the set of questions for peer reviewers in Section 7.4.
	Hold writing conference during which the reviewer returns the first draft to the writer.	An informal discussion between the writer and the reviewer is useful to give the writer an opportunity to explain what he/she intended to accomplish, and for the reviewer to provide feedback.
Days 6–7	Revise laboratory report.	Based on your discussion with your reviewer, revise as necessary. Remember that you do not have to accept all of the reviewer's suggestions.
Day 8	Hand in both first draft and revised draft to instructor.	Your instructor wants to know what you've learned (we never stop learning either!).

or thumb drive), an external hard drive, or online. See "Backing up your files" in Appendix 1 for benefits and drawbacks of various options.

TABLE 5.2 Instructions to authors of laboratory reports

Feature	Layout
Paper	8½" × 11" (or DIN A4) white bond
Margins	1.25" left and right; 1" top and bottom
Font size	12 pt (points to the inch)
Typeface	Times Roman or another *serif* font. A serif is a small stroke that embellishes the character at the top and bottom. The serifs create a strong horizontal emphasis, which helps the eye scan lines of text more easily.
Symbols	**Insert | Symbols** from the Word menu. Math AutoCorrect has pre-programmed keyboard shortcuts for many symbols (see Section A1.11 in Appendix 1).
Pagination	Arabic number, top right on each page except the first
Justification	Align left/ragged right or Full/even edges
Spacing	Double
New paragraph	Indent 0.5"
Title page (optional)	Title, authors (your name first, lab partner second), class, and date
Headings	Align headings for Abstract, Introduction, Materials and Methods, Results, Discussion, and References on left margin or center them. Use consistent format for capitalization. Do not start each section on a new page unless it works out that way coincidentally. Keep section heading and body together on the same page.
Subheadings	Use sparingly and maintain consistent format.
Tables and figures	Incorporate into text as close as possible after the paragraph where they are first mentioned. Use descriptive titles, sequential numbering, proper position above or below visual. May be attached on separate pages at end of document, but must still have proper caption. Keep table/figure and its caption together on the same page.
Sketches	Hand-drawn in pencil or ink. Other specifications as in "Tables and figures" above.
References	Citation-Sequence System: Make a numbered list in order of citation.
	Name-Year System: List references in alphabetical order by the first author's last name. Use a hanging indent (all lines but the first indented) to separate individual references.
	Both systems: Use accepted punctuation and format.
Electronic submission	Save file as a PDF. Upload to website.
Hard copy	Place pages in order, staple top left.

Save your file frequently while writing your paper by clicking 🖫 on the Quick Access Toolbar. You can also adjust the settings for automatically saving your file. Windows users, click **File | Options**. Mac users, click **Word | Preferences | Output and Sharing**. From there, the common sequence is **Save | Save AutoRecover information every __ minutes**.

Install antivirus software on your computer and always check flash drives for viruses before you use them. Beware of files attached to email messages. Do not open attachments unless you are sure they come from a reliable source.

Store flash drives with their caps on to keep dust out. Protect them from excess humidity, heat, and cold. Only remove a flash drive from a computer after you eject it and the message "Safe to Remove Hardware" is displayed.

If you must eat and drink near a computer, keep beverages and crumbs away from the hard drive and keyboard.

5.2 Getting Started

Set aside 1 hour to begin writing the laboratory report as soon as possible after doing the laboratory exercise. Turn off your phone and get off social networking sites. Writing lab reports requires your full concentration. What matters is the quality, not the quantity, of time you spend on your assignments. Promise yourself a reward for time well spent.

Reread the laboratory exercise

You cannot begin to write a paper without a sense of purpose. What were the objectives of your experiment or study? What questions are you supposed to answer? Take notes on the laboratory exercise to prevent problems with plagiarism when you write your laboratory report (see Sections 3.6 and 4.2).

Organization

If your instructor provided a rubric or other instructions for organizing your paper, follow the instructions exactly. Otherwise use the standard IMRD format, as described in Section 3.2.

Audience

Scientific papers are written for scientists. Similarly, laboratory reports should be written for an audience of fellow student-biologists, who have a knowledge base similar to your own. When deciding how much background information to include, assume that your audience knows only what you learned in class. Use scientific terminology, but define any terms or acronyms known only to experts (**jargon**).

Write for an audience of fellow scientists, not students in a classroom situation. Note the difference between the original text and the revision in the following examples:

FAULTY: The experiments performed by the students dealt with how different wavelengths of light affect seed germination.

REVISION: The purpose of the experiment was to determine how different wavelengths of light affect seed germination.

FAULTY: The purpose of this experiment is to become acquainted with equipment such as secchi disc, dissolved oxygen meter, pH meter, and Van Dorn bottle for sampling lake water.

REVISION: The purpose of this experiment was to assess the water quality in Lake Wallenpaupack.

Writing style

Laboratory reports are formal written assignments. Avoid slang and contractions and choose words that reflect the serious nature of scientific study. Readers of scientific papers trust the scientific method and are confident that the facts speak for themselves. For this reason, write objectively—that is, do not make judgments. When making a statement that may not be obvious to the audience, always back it up by citing an authoritative source or by providing experimental evidence. Because the focus is on the science, not the scientist, passive voice is used more frequently (especially in the Materials and Methods section) than in other kinds of writing. Use active voice in the other sections, however, because it makes sentences shorter and more dynamic.

Past and present tense have specific connotations in scientific papers. Authors use *present tense* to make *general statements* that the scientific community agrees are valid. Statements that are generally valid include explanations of phenomena based on experimental results that have been replicated by many scientists. Therefore, use present tense in the Introduction and Discussion sections when describing information accepted by the scientific community, and cite the source of any information that is not common knowledge for your audience. On the other hand, authors use *past tense* to make statements about *their own work*. For this reason, use past tense in the Materials and Methods and Results sections, and whenever you are describing work that you personally carried out.

5.3 Start with the Materials and Methods Section

The order in which you write the different sections is not the order in which they appear in the finished laboratory report. The rationale for this plan will become obvious as you read on. The Materials and Methods section requires the least amount of thought, because you are primarily restating the procedure in your own words.

Tense

When you write the Materials and Methods section, describe the procedure in *past*, not present, tense because (1) these are completed actions and (2) you are describing your own work. Do *not* copy the format of your laboratory exercise, in which the instructions may be arranged in a numbered list and the imperative (command) form of verbs may be used for clarity.

Voice

There are two grammatical voices in writing: active and passive. In active voice, the subject *performs* the action. In passive voice, the subject *receives* the action. Passive voice is preferred in the Materials and Methods section because the subject that receives the action is more important than who performed it. The logic is that anyone with the appropriate training should be able to perform the action. Consider the following examples:

> ACTIVE VOICE: I collected the water samples at two different depths.

> PASSIVE VOICE: Water samples were collected at two different depths.

The sentence written in active voice is more natural and dynamic, but it shifts the emphasis from the subject, "water samples," to "I." Passive voice places the emphasis on the water samples, where it belongs. Because sentences written in passive voice tend to be longer and less direct than those written in active voice, try to use active voice when the performer (you) is not the subject of the sentence.

Level of detail

A well-written Materials and Methods section will *provide enough detail to allow someone with appropriate training to repeat the procedure.* For example, for a **molecular biology** procedure, include essential details such as the concentration and pH of solutions, reaction and incubation times, volume, temperature, wavelength (set on a spectrophotometer), centrifugation speed, dependent and independent variables, and control and treatment groups. On the other hand, *do not describe routine lab procedures* such as:

- How to calculate molarity or use $C_1V_1 = C_2V_2$ to make solutions.
- Taring a balance before use.
- Using a vortex mixer to ensure that solutions are well mixed.
- Describing how to zero (blank) a spectrophotometer before measuring the absorbance of the samples.
- Explaining what type of serological pipette or micropipettor is appropriate for a particular volume.
- Designating the type of flasks or beakers to use.
- Specifying the duration of the entire study ("In our two-week experiment, …").

For a **field experiment**, however, time *is* important. When observing or collecting plants and animals in nature, be sure to include in the Materials and Methods section time of day, month, and year as appropriate; sampling frequency; location and dimensions of the study site; sample size; and statistical analyses. Depending on the focus of your lab report, it may also be prudent to describe the geology, vegetation, climate, natural history, and other characteristics of the study site that could influence the results.

Here are some guidelines for the level of detail to include in the Materials and Methods section.

Not enough information   Include all relevant information needed to repeat the experiment.

FAULTY: In this lab, we mixed varying amounts of BSA stock solution with varying amounts of TBS using a vortex mixer. We used a spectrophotometer to measure absorbance of the 4 BSA samples, and then we determined the concentration of 4 dilutions of egg white from the standard curve.

EXPLANATION: This procedure does not give the reader enough information to repeat the experiment, because essential details like *what concentrations of BSA* were used to construct the standard curve, *what dilutions of egg white* were tested, and the *wavelength* set on the spectrophotometer have been left out.

REVISION: Bovine serum albumin (BSA) solutions (2, 3, 5, 10 mg/mL) were prepared in tris-buffered saline (TBS). The egg white sample was serially diluted 1/5, 1/15, 1/60, and 1/300 with TBS. The absorbance of all samples was measured at 550 nm using a Genesys 30 spectrophotometer.

The following are examples of **too much information**.

Do not list materials and methods separately   The wording of the section heading makes it tempting to separate the content into two parts. In fact, materials should not be listed separately unless the strain of bacteria, vector (plasmid), growth media, or chemicals were obtained from a special or noncommercial source. It will be obvious to the reader what materials are required on reading the methods.

Describe the solutions, not the containers

FAULTY: Eight clean beakers were labeled with the following concentrations of hydrogen peroxide and those solutions were created and placed in the appropriate beaker: 0, 0.1, 0.2, 0.5, 0.8, 1.0, 5.0, and 10.0.

EXPLANATION: Using clean, suitable containers to store solutions is common practice in the laboratory. Putting labels on labware is also a routine procedure. An essential detail missing from this sentence is the units.

REVISION: The following hydrogen peroxide solutions were prepared: 0, 0.1, 0.2, 0.4, 0.8, 1.0, 5.0, and 10.0%.

Specify the concentrations, not the procedure for making solutions

FAULTY: To make the dilution, a micropipette was used to release 45, 90, 135, and 180 μL of bovine serum albumin (BSA) into four different test tubes. To complete the dilution, 255, 210, 165, and 120 μL of TBS was added, respectively.

EXPLANATION: With appropriate instruction, making dilutions of stock solutions becomes a routine procedure. In the above example, you should assume that your readers can make the solution using the appropriate measuring instruments *as long as you specify the final concentration.*

REVISION: The following concentrations of BSA were prepared for the Bradford assay: 300, 600, 900, and 1200 µL/mL.

Include only essential procedures and write concisely

FAULTY: The test tubes were carried over to the spectrophotometer and the wavelength was set to 595 nm (nanometer). The spectrophotometer was zeroed using the blank. Each of the remaining 8 samples in the test tubes were individually placed into the empty spec tube, which was then placed in the spectrophotometer where the absorbance was determined.

EXPLANATION: The only detail important enough to mention is the wavelength.

REVISION: The absorbance of each sample was measured with a Genesys 30 spectrophotometer at 595 nm.

Avoid giving "previews" of your data analysis

FAULTY: A graph was plotted with Absorbance on the *y*-axis and Protein concentration on the *x*-axis. An equation was found to fit the line, then the unknown protein absorbances that fell on the graph were plugged into the equation, and a concentration was found.

EXPLANATION: Making graphs is something that you do when you analyze your raw data, but it is not part of the experimental procedure. How and why you chose to plot the data will become obvious to the reader in the Results section, where you display graphs, tables, and other visuals and describe the noteworthy findings.

REVISION: Delete this entire passage.

Cite published sources   Published laboratory exercises must be cited (see Chapter 4). To cite unpublished material, see Section 4.5.

5.4 Do the Results Section Next

The Results section has two elements:

- **Visuals**, such as tables and figures, which display data. Visuals are numbered consecutively so that they can be described in order. Each visual is accompanied by a caption that consists of a number and an informative title.
- A **description** of the most important results shown in each visual

Preparing visuals

The most common visuals in scientific writing are tables and figures. A **table** is useful when it is important to have a record of the actual numbers or categories. A **figure** is any visual that is not a table. Thus, line graphs (also called XY graphs), bar graphs, pie graphs, drawings, gel images, microscope images, and phylogenetic trees are all called *figures* in scientific papers.

Graphs show relationships between or among variables. The type of graph that can be used is often dictated by the nature of the variables—quantitative or categorical. **Categorical variables** are groups or categories that have no units of measurement (treatment groups, age groups, habitat, etc.). Bar graphs and pie graphs are commonly used to display results involving categorical variables. **Quantitative variables**, on the other hand, have numerical values with units. Histograms show the distribution of one quantitative variable. XY graphs and scatter graphs (also called scatterplots) display relationships between two quantitative variables. Some of the graphs frequently encountered in the field of biology are summarized in Table 5.3 and described individually in the following sections.

Do not feel that you have to have visuals in your lab report. All or none results, for example, "none of the seeds germinated in buffered solutions less than pH 5," do not require a visual.

Tables

Tables are used to display large quantities of numbers and other information that would be tedious to read in prose. Arrange the categories vertically, rather than horizontally, as this arrangement is easier for the reader to follow (see, for example, Table 1 in Figure 5.1). List the items in a logical order (e.g., sequential, alphabetical, or increasing or decreasing value). Use sentence case for the headings. Include the units in each column heading to save yourself the trouble of writing the units after each number entry in the table.

By convention, tables in scientific papers do not have vertical lines to separate the columns, and horizontal lines are used only to separate

TABLE 5.3	Types of graphs and their purpose	
Graph	Purpose	Example
Histogram	To show the distribution of a quantitative variable.	Distribution of grades on an exam. Y-axis shows number of students; x-axis shows numerical score on the exam.
Scatterplot	To show the relationship between two quantitative variables measured on the same individuals. Look for an overall pattern and for deviations from that pattern. If the points lie close to a straight line, a linear trendline may be superimposed on the scatter graph. The correlation, r, indicates the strength of the linear relationship.	Relationship between shell length and mass. If we are just looking for a pattern, it doesn't matter which variable is plotted on which axis. If we suspect that mass depends on length, plot mass on the y-axis and length on the x-axis. Look at the form, direction, and strength of the relationship.
XY graph	To show the relationship between two quantitative variables. One variable may be dependent on the other. The variable that is being manipulated is called the independent or explanatory variable. The variable that changes in response to the independent variable is called the dependent or response variable. By convention, the independent variable is plotted on the x-axis and the dependent variable is plotted on the y-axis. Error bars may be included to show variability.	Relationship between enzyme activity and temperature. Because temperature is the variable that is being manipulated, it is plotted on the x-axis. Because enzyme activity is the response being measured, it is plotted on the y-axis.
Scatterplot with regression line	To predict the value of y for a given value of x or vice versa. The response variable must be dependent on the explanatory variable and the relationship must be linear. The regression line takes the form $y = mx + b$, where m is the slope and b is the y-intercept.	Standard curve for a protein assay. Protein concentrations of a standard such as BSA are plotted on the x-axis. Absorbance (measured by a spectrophotometer) for each concentration is plotted on the y-axis. A regression line is fitted to the data. To predict the protein concentration of a sample (x), measure its absorbance (y) and solve the regression equation for x.
Bar graph	To show the distribution of a categorical (non-quantitative) variable.	Effect of different treatments on plant height. One axis shows the treatment category and the other shows the numerical response.
Pie graph	To show the distribution of a categorical (non-quantitative) variable in relation to the whole. All categories must be accounted for so that the pie wedges total 100%.	Composition of insects in a backyard survey. Each wedge represents the percentage of an order of insects. Orders with low representation may be combined into an "Other" wedge to complete the pie.

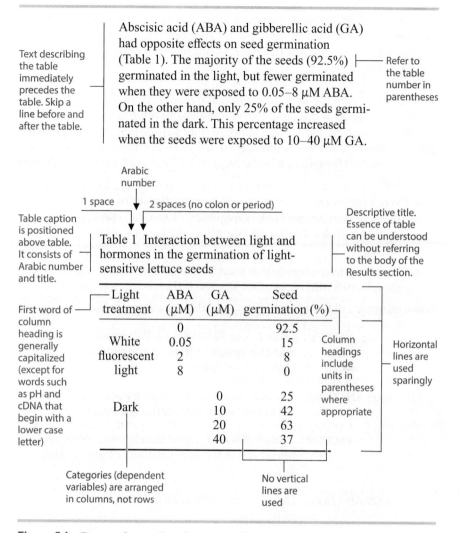

Figure 5.1 Excerpt from a Results section showing a properly formatted table preceded by the text that describes the data in the table.

the table caption from the column headings, the headings from the data, and the data from any footnotes. The tables in this book are formatted in this style.

Tables can be constructed in either Microsoft Word (see Section A1.12 in Appendix 1) or Microsoft Excel (see Section A2.2 in Appendix 2).

Table captions Give each table a caption that includes a number and a title. Center the caption or align it on the left margin *above* the table. Use Arabic numbers, and number the tables consecutively in the order they are discussed in the text. Notice that in this book, the table and figure numbers are preceded by the chapter number. This system helps orient the reader in long manuscripts, but is not necessary in short papers like your laboratory report.

Table titles   From the table title alone, the reader should be able to understand the essence of the table without having to refer to the body (text) of the Results section. For simple tables, it may suffice to use a precise noun phrase rather than a full sentence for the title. For more complex tables, one or more full sentences may be required. Either way, English grammar rules apply:

- Do not capitalize common nouns (*general* classes of people, places, or things) unless they begin the phrase or sentence.
- Capitalize proper nouns (names of *specific* people, places, or things).
- Do not capitalize words that start with a lowercase letter (for example, pH, mRNA, or cDNA), even if they begin a sentence.

Some examples of faulty and preferred titles are shown below.

FAULTY: Table 1 The Relationship Between Light and Hormones in the Germination of Light-Sensitive Lettuce Seeds

EXPLANATION: Use sentence case. Do not capitalize common nouns unless they start the sentence.

FAULTY: Table 1 Table of interaction between light and hormones in the germination of light-sensitive lettuce seeds

EXPLANATION: Do not start a title with a description of the visual.

FAULTY: Table 1 Seed germination data

EXPLANATION: Do not write vague and undescriptive titles.

REVISION: Table 1 Interaction between light and hormones in the germination of light-sensitive lettuce seeds

In your laboratory report, it is not necessary to include a table when you already have a graph that shows the same data. Make *either* a table *or* a graph—not both—to present a given data set.

Table preparation checklist

❏ Categories arranged in columns, not rows
❏ Column headings include units (where appropriate)
❏ Format correct (minimal lines)
❏ Table title descriptive
❏ Table title in sentence case
❏ Table caption positioned above the table

XY graphs and scatterplots

XY graphs and scatterplots display a relationship between **two or more quantitative variables**. For experimental data, the **independent variable** (the one the scientist manipulates) is plotted on the x-axis and the **dependent variable** (the one that changes in response to the independent variable) is plotted on the y-axis. **Time series plots** are a special type of XY graph in which time is one of the variables. In time plots, time is always plotted on the x-axis, and the measured variable is plotted on the y-axis. Instructions for plotting XY graphs in Excel 2019 and Excel for Mac 2019 are given in Section A2.4 in Appendix 2.

Lines For experimental data, usually some kind of line or curve is superimposed on the individual data points to emphasize *how* the dependent variable changes with the independent variable. How do you decide what kind of line or curve to insert? Figure 5.2 gives some examples.

- **Scatterplots** are graphs in which the individual points are **not connected**. These kinds of graphs are often used for data exploration, to determine the form, direction, and strength of the relationship between two variables. Data exploration is usually done before statistical hypothesis testing (see Section 6.5).
- When plotting measured data for which there is **no known mathematical relationship**, connect the points with **straight lines**. Straight lines signal that you are not inferring relationships about values that you didn't measure. On the other hand, if the in-between values can be interpolated with confidence, it may be appropriate to insert a **smoothed** line.
- When there is an **expected mathematical relationship** between the variables, insert the appropriate **trendline**. For example, standard curves (calibration curves) based on Beer's Law tend to be **linear**. Enzymatic activity increases **logarithmically** as a function of substrate concentration. Bacterial growth curves have an **exponential** phase. Knowing the mathematical function allows you

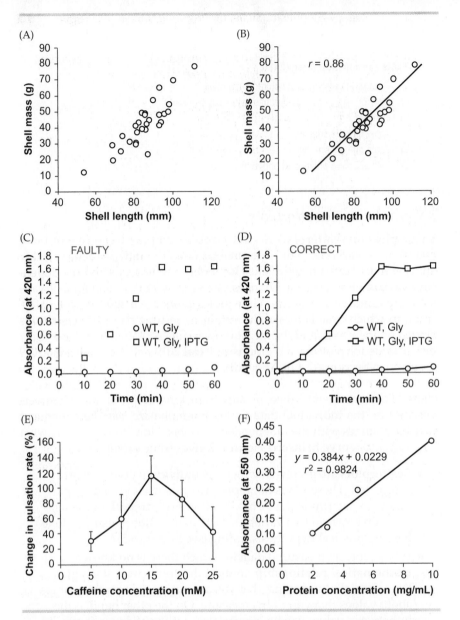

Figure 5.2 Different XY graph formats showing relationships between variables. (A) A scatterplot displays numerical observations with the purpose of determining whether there is a relationship between shell mass and shell length. (B) A scatterplot with a straight line added shows that there is a strong linear relationship between the two variables. (C) The relationship between the independent and dependent variables in each treatment group is hard to see when the points are not connected. (D) The relationship is much easier to see when the points are connected with straight or smoothed lines. (E) Error bars show variability about the mean. (F) A least-squares regression line and its equation are used to predict one variable when the other is known. Linear regression lines are used only in specific situations when the mathematical relationship between the two variables is clearly established.

to make predictions. For example, a standard curve for a protein assay lets you predict the unknown protein concentration of a sample from its measured absorbance.

- Is it ever appropriate to plot a line *without* the measured data points? No. If values were measured in the study, they must be displayed because they provide experimental evidence that makes it possible to draw conclusions about the relationship between the variables.

Correlation versus causation Correlation is a statistical measure of the relationship between two variables. However, correlation does not imply that one variable causes the other. This distinction is especially important in observational studies, where the scientist cannot control all of the variables that might affect a certain outcome. For example, an observational study might show a positive association between overweight people and those who use artificial sweeteners. Some might argue that these results show that artificial sweeteners *cause* weight gain. But there may be alternative explanations. Perhaps overweight people are choosing sugar substitutes, because they are trying to lose weight by reducing their sugar consumption. This would explain the correlation, but would not imply a causal mechanism between artificial sweeteners and weight gain. Or perhaps overweight people are experiencing cravings for sweets, which prevent them from losing weight even though they are using artificial sweeteners. In this case, the causal mechanism could be consumption of other sources of sugar.

A better way to demonstrate a causal mechanism between variables is to carry out a carefully controlled experiment. In any given experiment, scientists change only one variable, while keeping all other variables constant. This approach makes it more likely that any changes in the dependent variable are caused by changes in the independent variable. Causation is especially difficult to prove in experiments involving humans (clinical trials), because of the influence of heredity and human behavior.

Figure captions Figures are always numbered and titled *beneath* the visual (Figure 5.3). The captions may be centered or placed flush on the left margin of the report. Arabic numbers are used, and the figures are numbered consecutively in the order they are discussed in the text.

Figure titles From the figure title alone, the reader should be able to understand the essence of the figure without having to refer to the body (text) of the Results section. For simple figures, it may suffice to use a precise noun phrase rather than a full sentence for the title. For more complex figures, one or more full sentences may be required. Either way, English grammar rules apply: Do not capitalize common nouns (*general* classes of people, places, or things) unless they begin the phrase or sentence. Capitalize proper nouns (names of *specific* people, places, or things). Do

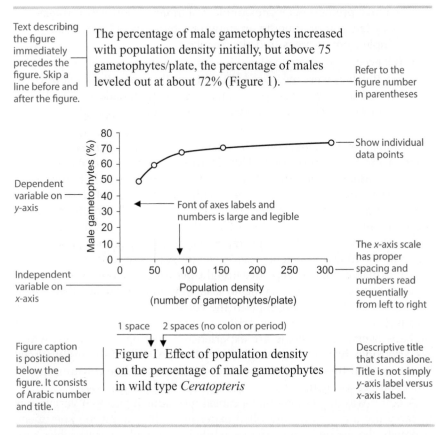

Text describing the figure immediately precedes the figure. Skip a line before and after the figure.

The percentage of male gametophytes increased with population density initially, but above 75 gametophytes/plate, the percentage of males leveled out at about 72% (Figure 1).

Refer to the figure number in parentheses

Show individual data points

Dependent variable on y-axis

Font of axes labels and numbers is large and legible

The x-axis scale has proper spacing and numbers read sequentially from left to right

Independent variable on x-axis

1 space   2 spaces (no colon or period)

Figure caption is positioned below the figure. It consists of Arabic number and title.

Figure 1 Effect of population density on the percentage of male gametophytes in wild type *Ceratopteris*

Descriptive title that stands alone. Title is not simply y-axis label versus x-axis label.

Figure 5.3 Excerpt from a Results section showing a properly formatted figure with one line (data set); text that describes the figure precedes it.

not capitalize words that start with a lower case letter (for example, pH, mRNA, or cDNA), even if they begin a sentence. Some examples of faulty and preferred titles for the graph in Figure 5.3 are shown here.

FAULTY: Figure 1 The Effect of Population Density on the Development of Male Gametophytes

EXPLANATION: Use sentence case. Do not capitalize common nouns unless they start the sentence.

FAULTY: Figure 1 Percentage of male gametophytes vs. population density

EXPLANATION: Do not restate the y-axis label versus the x-axis label as the figure title.

FAULTY: Figure 1 shows the effect of population density on the percentage of male gametophytes in wild type *Ceratopteris*

EXPLANATION: Separate the figure number and the title.

FAULTY: Figure 1 Line graph of the effect of population density on the percentage of male gametophytes in wild type *Ceratopteris*

EXPLANATION: Do not start a title with a description of the visual.

FAULTY: Figure 1 Averaged class data for C-fern experiment

EXPLANATION: Do not write vague and undescriptive titles.

REVISION: Figure 1 Effect of population density on the percentage of male gametophytes in wild type *Ceratopteris*

More than one data set   When there is more than one data set (line) on the figure, you have three options:

- Add a brief label (no border, no arrows) next to each line.
- Use a different symbol for each line and label the symbols in a legend (as in Figure 5.4). Place the legend without a border within the axes of the graph. This is the easiest option if you are using Excel to plot your data.
- If the first two options make the figure look cluttered, identify the symbols in the figure caption.

All three formats are acceptable in scientific papers as long as you use them consistently.

Black and white or color   Black lines, text, numbers, and symbols give the best contrast on a white background. Black is preferred for *printed* manuscripts because, when they are copied or printed on a black and white printer, there is no ambiguity about black on white. Colored lines turn out to be various shades of gray, which may make it hard for the reader to distinguish the numbers on the axes and the different data sets.

Online journals routinely publish in color, as there are no additional costs involved. If you submit your lab reports electronically, it's a good idea to check with your instructor about their policy on using color. If color is allowed, choose colors wisely to make your figures accessible to colorblind

The concentration of cells in both cultures remained low for several hours (Figure 2). At about 4 hr, the concentration of the ABC strain increased rapidly and then leveled off after about 9 hr. On the other hand, it took longer for the concentration of strain XYZ to increase, but at 12 hr, the concentration was much higher than that of strain ABC.

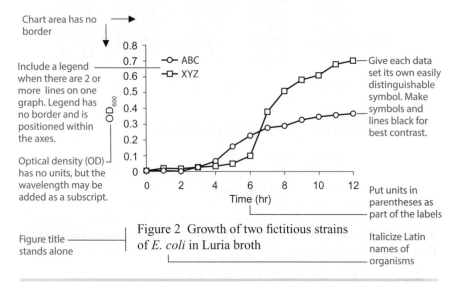

Chart area has no border
Include a legend when there are 2 or more lines on one graph. Legend has no border and is positioned within the axes.
Optical density (OD) has no units, but the wavelength may be added as a subscript.
OD_{600}
0.8
0.7
0.6
0.5
0.4
0.3
0.2
0.1
0
ABC
XYZ
0
2
4
6
8
10
12
Time (hr)
Give each data set its own easily distinguishable symbol. Make symbols and lines black for best contrast.
Put units in parentheses as part of the labels
Figure title stands alone
Figure 2 Growth of two fictitious strains of *E. coli* in Luria broth
Italicize Latin names of organisms

Figure 5.4 Excerpt from a Results section showing a properly formatted figure with two sets of data. A legend (key) is needed to distinguish the two lines. The text that describes the figure precedes it.

people. Using different-shaped symbols *and* different line patterns increases the chances that your reader will be able to distinguish the data sets easily.

Bar graphs

A bar graph allows you to compare individual sets of data when **one of the variables is categorical** (not quantitative)—this is the main difference between XY graphs and bar graphs. Bar graphs are more flexible than pie charts because any number of categories can be compared; the percentages do not have to total 100%.

Consider an experiment in which you want to compare the final height of the same species of plant treated with four different nutrient solutions. The nutrient solution is the non-numerical, categorical variable; the height is the response variable. The data bars are arranged vertically in Figure 5.5, because the category labels are short.

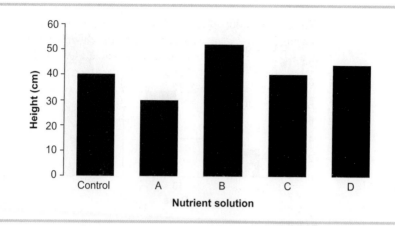

Figure 5.5 Final height of corn plants after 4-week treatment with different nutrient solutions. This figure is an example of a column graph.

Figure 5.6, on the other hand, is an example of a bar graph with horizontal bars. This arrangement is more convenient to accommodate the long categorical labels that describe the rats' maternal diet and diet after weaning. The dependent variable is the time the rats spent searching for the hidden escape platform in the memory retention portion of the Morris water maze test.

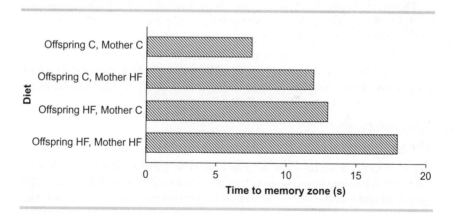

Figure 5.6 The time male rats spent swimming to the location of a submerged platform in the probe test portion of the Morris water maze. The study was conducted to determine the effect of maternal and post-weaning diet on memory retention in rats. HF = high fat diet, C = control diet. Data kindly provided by Professor Kathleen Page, Bucknell University. This figure is an example of a horizontal bar graph with long category labels and pattern bars.

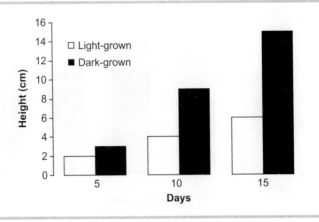

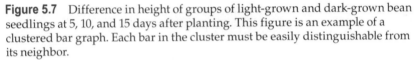

Figure 5.7   Difference in height of groups of light-grown and dark-grown bean seedlings at 5, 10, and 15 days after planting. This figure is an example of a clustered bar graph. Each bar in the cluster must be easily distinguishable from its neighbor.

The bars should be placed sequentially, but if there is no particular order, then put the control treatment bar far left in column graphs or at the top in horizontal bar graphs. Order the experimental treatment bars from shortest to longest (or vice versa) to facilitate comparison among the different conditions. The baseline does not have to be visible, but all the bars must be aligned as if there were a baseline.

The bars should always be wider than the spaces between them. In a graph with clustered bars, make sure each bar has sufficient contrast so that it can be distinguished from its neighbor (Figure 5.7).

Gridlines may make it easier to read the value of the bars. If gridlines are used, they should be unobtrusive (thinner or lighter than the axes). Instructions for plotting bar graphs in Excel 2019 and Excel for Mac 2019 are given in Section A2.4 in Appendix 2.

Pie graphs

A pie graph is used to show data as a percentage of the total data. For example, if you were doing a survey of insects found in your backyard, a pie graph would be effective in showing the percentage of each kind of insect out of all the insects sampled (Figure 5.8). There should be between two and eight segments in the pie. Place the largest segment in the right-hand quadrant with the segments decreasing in size clockwise. Combine small segments under the heading "Other." Position labels and percentages horizontally outside of the segments for easy reference. Instructions for

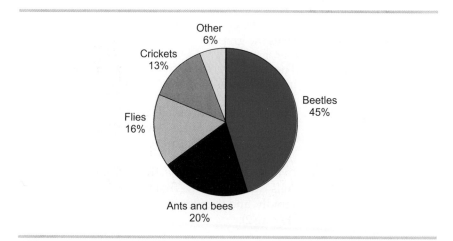

Figure 5.8 Composition of insects in backyard survey. Pie graphs are used to show data as a percentage of the total data.

plotting pie charts in Excel 2019 and Excel for Mac 2019 are given in Section A2.4 in Appendix 2.

Figure preparation checklist
❑ Right type of graph
❑ Format correct (symbols, lines, legend, axis scale, outside tick marks, no gridlines (but see "Bar graphs"), no border)
❑ Figure title descriptive
❑ Figure title in sentence case
❑ Figure caption positioned below the figure

Writing the body of the Results section

Now that your data are displayed visually in graphs or tables, it's time to tell your reader what you consider to be important. The suggestions below will help you get started.

Use words that describe relationships that the reader can readily see in the figure How does the dependent variable change with the independent variable? Is there a recognizable mathematical relationship (for example, linear, logarithmic, or exponential) or does the trend require a more detailed description?

For example, Figure 5.4 shows an excerpt from a Results section. In this experiment, the growth of two fictitious strains of bacteria was monitored over time. The dependent variable, optical density (OD_{600}), is a measure of the concentration of cells in the culture. Consider how well each of the following sentences describes the graph.

> VAGUE: The concentration of cells increased over time in both strains of *E. coli* (Figure 2).

> BETTER: The concentration of cells remained low at first and then increased over time in both strains of *E. coli* (Figure 2).

> BETTER YET: In both strains of *E. coli*, the concentration of cells was low for the first few hours, increased rapidly, and then leveled off after about 12 hr (Figure 2).

How much detail you use to describe the shape of the curve depends on how you plan to explain the results in the Discussion section. Pointing out key features in the Results section prepares the reader for what is to come in the discussion. The discussion is where you would correlate the changes in slope of your graph with the lag, log, and stationary phases of bacterial growth.

When there is more than one data set or category on a figure, make comparisons

> TEDIOUS: In the ABC strain, concentration was low for the first few hours, increased rapidly after 4 hr, and then leveled off (Figure 2). In the XYZ strain, concentration was low for the first few hours, increased rapidly after 6 hr, and then leveled off.

> BETTER: In both strains of *E. coli*, the concentration of cells was low for the first few hours, increased rapidly, and then leveled off (Figure 2). The concentration in the ABC culture began to increase after 4 hr and reached a maximum after about 12 hours. In contrast, the concentration of XYZ cells started to increase later, but reached a higher level after 12 hours.

Don't explain or interpret the results

> FAULTY: Both strains of *E. coli* displayed a lag phase, followed by a log growth phase in which the cells divided rapidly, and then a stationary

phase in which the growth rate was counterbalanced by the death rate (Figure 2).

EXPLANATION: The terms *lag phase, log growth phase*, and *stationary phase* are not marked on the graph, nor can the reader tell from the graph what the cells are doing. Save interpretations like this for the Discussion section.

Refer to the figure that shows the result you are describing

FAULTY: In both strains of *E. coli*, concentration was low for the first few hours, increased rapidly, and then leveled off.

EXPLANATION: Tell the reader where to find the data. Reference the figure in parentheses at the end of the sentence.

FIGURE AS SUBJECT: Figure 2 shows that in both strains of *E. coli*, concentration was low for the first few hours, increased rapidly, and then leveled off.

PREFERRED: In both strains of *E. coli*, concentration was low for the first few hours, increased rapidly, and then leveled off (Figure 2).

It is not incorrect to make the figure reference the subject of the sentence. However, notice how this style places the emphasis on the figure, instead of the results. The preferred style emphasizes the results, which is what the reader is looking for.

When you need more than one sentence to describe the results in a visual, refer to the figure just once in that paragraph, preferably at the beginning. That way, your readers will know right away where to look for the data, and they will assume you are describing the same figure unless you tell them otherwise.

Some XY graphs are intended to be used as tools, not to show relationships For example, standard curves (calibration curves) are specific types of XY graphs that show absorbance as a function of concentration. Their purpose is not to show that absorbance is proportional to concentration (Beer's Law already establishes that fact), but to allow the concentration of an unknown to be calculated from its measured absorbance. When writing about standard curves, don't describe the trend. Instead, describe how the graph was used to calculate the unknown parameter of interest.

Make every sentence meaningful

FAULTY: Concentration changed over time in both strains of *E. coli* (Figure 2).

EXPLANATION: This sentence does not say *how* concentration changed over time.

FAULTY: The results in Figure 2 show the averaged data for the whole class.

EXPLANATION: This sentence does not describe an actual result. Describe statistical methods in the Materials and Methods section. Alternatively, state that these are averaged data, and include the number of trials in the figure caption.

FAULTY: After the results were obtained, a graph was made with time on the *x*-axis and concentration on the *y*-axis, as shown in Figure 2.

EXPLANATION: This sentence states the obvious and should be deleted. Describe results, not axis labels.

Eliminate unnecessary introductory phrases. This includes phrases such as

- It was found that…
- The results showed that…
- It could be determined that…

Get to the point! State the important results in clear, concise terms.

REVISION: Delete the introductory phrase and begin the sentence with an actual result.

Use past tense   Whenever you are referring specifically to your own results, use past tense. In scientific papers, present tense is reserved for statements accepted by the scientific community as fact. At this point, your results are not yet considered "fact."

Final layout

Your results section will typically consist of text alternating with visuals (figures and tables). The idea is to describe the visual first, reference it by number, and then display the visual (Figure 5.9) This arrangement prepares the reader for what to look for before seeing the visual. Some

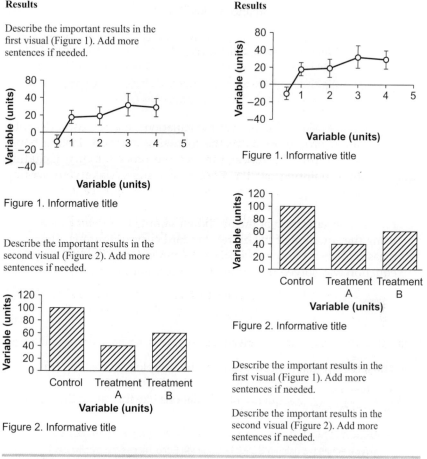

Figure 5.9 Layout of text and figures in a Results section. (A) The text describing the results in a figure precedes the figure, so that text and figures alternate in this section. (B) This is not an acceptable layout for the Results section.

journals require figures and tables to be submitted as an appendix or in separate files. Please confirm the required format for your lab report with your instructor.

Equations

Equations are technically part of the text and should *not* be referred to as figures. Equations are set off from the rest of the text on a separate line. If you have several equations and need to refer to them unambiguously in the body of the Results (or another) section, number each equation sequentially and place the number in parentheses on the right margin. For example:

$$\text{Absorbance} = -\log T \qquad\qquad (1)$$

In Microsoft Word, to center an equation and right-align the equation number, insert a center tab stop and a right tab stop as shown.

When you are asked to show your calculations, use words to describe your calculation procedure, as in the following example. Listing equations without guiding the reader through the process is like including figures without pointing out the important results. Make your writing reader-friendly!

Protein concentration of the unknown sample was determined using the equation of the biuret standard curve (Figure 1). The measured absorbance value was substituted for y, and the equation was solved for x (the protein concentration):

$$y = 0.0417x$$

$$0.225 = 0.0417x$$

$$5.40 = x$$

Thus, the protein concentration of the sample was 5.40 mg/mL.

When you present a sequence of calculations like this, align the = symbol in each line.

Type equations using MS Word's Equation Editor, accessed by clicking **Equation** on the **Insert** tab. Type the first equation into the box. Press **Enter**. Repeat the process for each equation in the group. To align the group of equations on the equal sign, select all of the equations, right-click, and select **Align at =**. This method does not work if each line has a right-aligned equation number. In that case, the equations have to be aligned manually.

5.5 Make Connections

Now that the core of your report is done, it's time to describe how your work fits into the existing body of knowledge. These connections are made in the Discussion and Introduction sections. Ideally, there will be a one-to-one correspondence between the objectives stated in the Introduction and the interpretation and explanation of the results in the Discussion.

Write the discussion

The Discussion section gives you the opportunity to **interpret your results, relate them to published findings, and explain why they are important**. The structure of the discussion is specific to broad, as illustrated by the following triangle.

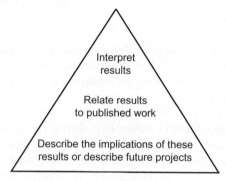

Interpret
results

Relate results
to published work

Describe the implications of these
results or describe future projects

A springboard   How do you start a Discussion section? There is no specific formula, but here are some suggestions.

- Summarize your most important result and then explain what it means.
- State whether your hypothesis was negated or supported and then provide the evidence.
- Reference a published journal article and state whether or not your results supported those findings.

If, initially, you don't know what your results mean, consult your resources. Does your lab handout give you any clues? Did your instructor talk about this topic in lecture? Don't forget to check the index of your textbook. When you think you have an explanation, make sure you use your own words and cite your sources. No matter how you start, by the time you finish your discussion, you should have interpreted your results in the context of your objectives and the work of other scientists.

When results defy explanation   Especially in introductory biology labs, the results may not always work out the way we expect. If your results defy explanation, consider these possible reasons:

- Human error, including failure to follow the procedure, failure to use the equipment properly, failure to prepare solutions correctly, variability when multiple lab partners measure the same thing, and simple arithmetic errors. If you suspect that human error negatively affected your results, then say so, but don't waste a lot of words in doing so.
- Numerical values were entered incorrectly in the computer plotting program. If possible, correct the errors and repeat the analysis.
- Sample size was too small. If possible, collect more samples.
- Variability was too great to draw any conclusions. Consider redesigning your experiment.

If you can rule out these possibilities, discuss your results with your teaching assistant or instructor. Having a discussion with a knowledgeable individual may help you better understand the concepts, even if your results didn't turn out the way you expected.

Results never prove hypotheses   When explaining your results, never use the word *prove*. Instead, use words and phrases like *provide evidence for, support, indicate, demonstrate*, or *strongly suggest*. The reason for this choice of words lies in the logic behind the scientific method. If our results match our predictions, then there is evidence that our hypotheses are correct. When many scientists get the same results independently, then the support for a given hypothesis grows. Scientists are reluctant to use the word "prove," because there is always a chance that a future study may provide conflicting evidence.

FAULTY: These results prove that catalase was denatured at temperatures above 60 °C.

REVISION: These results strongly suggest that catalase was denatured at temperatures above 60 °C.

Build your case   Writing a Discussion section is an exercise in persuasive writing. You want to convince readers that your conclusions are valid. To do so requires substantiating statements with experimental evidence and referencing the work of other researchers.

To illustrate this approach, let's look at a study on the effect of human activities on the biodiversity of gastropods in rocky intertidal areas in southern California (Roy et al. 2003)*. The authors suggested that tram-

---

* Roy K, Collins AG, Becker BJ, Begovic E, Engle JM. 2003. Anthropogenic impacts and historical decline in body size of rocky intertidal gastropods in southern California. *Ecology Letters* 6: 205–211.

pling, shell collecting, and harvesting of these mollusks for food or bait caused a decrease in body size over time.

Component	Rationale
Statement: Adults collected prior to 1960 were larger than those surveyed more recently.	Relates back to an observation in the Introduction that humans tend to harvest or collect the larger specimens
Data presented in Figure 2: Museum specimens collected before 1960 were significantly larger than those collected between 1961 and 1980 and those surveyed in the field.	Results support first statement
Data presented in Table 1: There was a decrease in median size and in the size of the largest individuals.	Results provide additional support for first statement
Counterargument: Another factor, such as climate change, could be responsible for the decrease in size.	Other reasons for results were considered
Way to test counterargument: If decrease in body size is due to human activities, then study sites in which human activity is prohibited should have larger sized mollusks.	A testable hypothesis is proposed for differentiating the effect of human activities and non-human activities
Data presented in Figure 3: Protected sites had larger individuals.	Results support human activities hypothesis
Comparison with other studies: Similar results were found with other species in CA.	3 published articles are cited
Comparison with other studies: Similar results were found worldwide.	5 published articles are cited
Future research: Investigate all possible reasons why large gastropods are disappearing from the southern CA coast; investigate ways to reverse negative human impact	The authors recognize that there may be more than one reason why body size is declining. Future work should test alternative hypotheses.

To test this hypothesis, the authors examined museum shell collections dating back to the late 1800s, and they also surveyed rocky intertidal sites where some of the museum specimens had been collected. Some of the survey sites were protected from human activities; at those sites, collection of invertebrates was prohibited.

In summary, this discussion is well organized, and the authors provide compelling evidence for their conclusions. Potentially controversial statements are supported with experimental results. Counterarguments were considered and refuted. References to other published papers enhance the credibility of this study.

Tense   When *describing* your own results, use *past* tense. However, when you use scientific fact to *explain* your results, use *present* tense.

> PAST TENSE: The initial velocity of the reaction *was* zero at temperatures between 60 °C and 90 °C (Figure 1).

> EXPLANATION: Past tense signifies that you are describing your own results.

> PRESENT TENSE: At high temperatures, there is no enzymatic activity because the enzymes *are* denatured.

> EXPLANATION: Present tense signifies that these statements are generally valid and considered to be scientific fact.

Compare your results with those in the literature   Do your results agree with those in published papers? If so, then your work supports the existing body of knowledge. If not, could a different method account for a conflicting result? If warranted, discuss possible reasons why your results did not turn out as expected.

Describe future work   You might propose hypotheses and experiments that build on your results.

Write the introduction

After having written drafts of the Materials and Methods, Results, and Discussion sections, you should be intimately familiar with the procedure, the results, and what the results mean. Now you are in a position to put your study or experiment into perspective. What was already known about the topic? Were there any unanswered questions? Why did you carry out this work?

The structure of the introduction is broad to specific, just the opposite of that of the discussion.

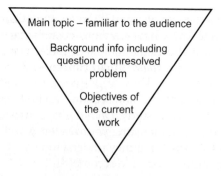

Main topic – familiar to the audience

Background info including question or unresolved problem

Objectives of the current work

Organization   The introduction consists of two main parts:

- Background information from the literature, and
- Objectives of the current work.

The opening sentence of the Introduction section is usually a **general observation or result** familiar to readers in that discipline. The author then quickly **narrows down the topic** and **provides background information** from the literature. The author then sets the stage for the current study by stating **any unanswered questions** or inconsistencies with previous work. Finally, in the last paragraph of the Introduction section, the author states the **objectives of the current work**. Specific hypotheses may be included if the study lends itself to hypothesis testing.

Start at the end   Deciding how much background information to include is a daunting task. To make this task slightly less daunting, **write your introduction in reverse order**. In other words, **start with the objectives** and then gradually fill in the *minimum* amount of information your reader would need to understand why you chose those particular objectives. You may already have directly or indirectly addressed your objectives in the Discussion section. In that case, restate those sentences in the Introduction section. Alternatively, get inspiration from looking at the variables on the figures in the Results section. XY graphs typically show the effect of the x-axis variable on the y-axis variable. Photographs are used to show relationships between form and function. Flow diagrams illustrate processes. Gel images show bands that represent the size and amount of a particular nucleic acid or protein. Maps show distributions. Phylogenetic trees show evolutionary relationships. By studying the relationships, patterns, and structures shown in the figures and tables, you should be able to identify the specific objectives of your study.

Now work backward from your objectives. **What concepts would the reader need to be reminded of for these objectives (or specific hypotheses) to make sense?** For example, if you set out to determine how temperature affects the rate of a particular enzyme-substrate reaction, the reader would need to know how enzyme and substrate molecules interact and how temperature affects the motion of molecules. **How much detail should you include about these broad concepts?** Less is better. In other words, provide just enough detail to prepare your readers for what is to come in the Results and Discussion sections. Include only the details that are directly relevant to your study. Do not give exhaustive reviews of the topic, otherwise you risk exhausting your reader! A well written introduction leaves readers satisfied that they understand why the experiment was done and what the author hoped to accomplish.

When you provide background information, be sure to **cite your sources**, especially when the information is not common knowledge (to determine what is common knowledge, see Section 4.1). Citing sources not only makes your statements more credible, it allows readers to find additional information in a related paper.

Finally, write the opening sentence. This may be the most difficult part of the introduction to write. To help you do so, focus on your readers. Who are they and what are they likely to know about this topic? Come up with several opening sentences and then evaluate each sentence critically in terms of the level of your audience. Choose an opening sentence that is neither too simplistic nor too technical.

Tense   In the course of providing background information on your topic, you will discuss scientific fact that is based on findings published in research papers. When describing scientific fact, use *present* tense.

> FACT: Peroxidase *is* completely denatured at temperatures above 80 °C (Duarte-Vázquez *et al.* 2003).

On the other hand, when stating the objectives of your study, use *past* tense. Past tense is preferred because proposing objectives is a completed action that you carried out before starting your actual study.

> OBJECTIVES: The purpose of this experiment *was* to determine the effect of temperature on peroxidase activity.

If the experiment lends itself to hypothesis testing, then state your hypothesis using a mixture of tenses. Notice how tenses are used in the following examples.

> HYPOTHESIS: We hypothesized [past tense] that enzyme activity will increase [future tense] with temperature up to a point.

HYPOTHESIS: We expected [past tense] enzyme activity to increase [present tense] with temperature up to a point.

HYPOTHESIS: Enzyme activity is or was [either present or past tense is appropriate] expected to increase with temperature up to a point.

Voice Active voice is preferred because it makes sentences shorter and more direct. But voice can also change the emphasis of a sentence, as illustrated by the following examples:

ACTIVE VOICE: Human activities are threatening the extinction of many species.

PASSIVE VOICE: The extinction of many species is threatened by human activities.

If your point is to emphasize *the role of human activities* in species extinctions, then active voice makes the stronger statement. If your focus is on *species extinction*, then passive voice may be more appropriate.

5.6 Effective Advertising

The whole point of writing your paper is to communicate your work to your fellow scientists. The abstract and the title are the primary tools potential readers will use to decide whether or not they are interested in your work.

Write the abstract

The abstract is a **summary of the entire paper** in 250 words or less. It contains:

- An introduction (scope and purpose)
- A short description of the methods
- The results
- Your conclusions

There are no literature citations or references to figures in the abstract.

After the title, the abstract is the most important part of the scientific paper used by readers to determine initial interest in the author's work. Abstracts are indexed in databases that catalogue the literature in the biological sciences. If an abstract suggests that the author's work may be relevant to your own work, you will probably want to read the whole article. On the other hand, if an abstract is vague or essential information is missing, you will probably decide that the paper is not worth reading. When you write the abstract for your own laboratory report or research paper, put yourself in the position of the reader. If you want the reader to be interested in your work, write an effective abstract.

Writing the abstract is difficult because you have to condense your entire paper into 250 words or less. One strategy for doing this is to list the key points of each section, as though you were taking notes on your own paper. Then write the key points in full sentences. Revise the draft for clarity and conciseness using strategies such as using active voice, combining choppy sentences with connecting words, rewording run-on sentences, and eliminating redundancy. With each revision, look for ways to shorten the text so that the resulting abstract is a concise and accurate summary of your work.

The ability to write abstracts is important to a scientist's career. Should you someday wish to present your research at an academic society meeting, such as the Society for Neuroscience, the American Association for the Advancement of Science, or the National Association of Biology Teachers (to name just a few), you will be asked to submit an abstract of your presentation to the committee in charge of the meeting program. Your chances of being among the select field of presenters at these meetings are much better if you have learned to write a clear and intelligent abstract.

Write the title

The title is a **short, informative description** of the essence of the paper. You may choose a working title when you begin to write your paper, but revise the title after subsequent drafts. Remember that readers use the title to determine initial interest in the paper, so descriptive accuracy is the most essential element of your title. Brevity is nice if it can be achieved. Some journals (especially the British ones) are fond of puns and humor in their titles, but this kind of thing may be better left for later in your career.

Here are some examples of vague and undescriptive titles:

FAULTY: Quantitative Protein Analysis

FAULTY: The Assessment of Protein Content in an Unknown Sample

FAULTY: Egg White Protein Analysis

EXPLANATION: These titles leave the reader wondering what method of protein analysis was used and what sample was analyzed.

REVISION: Measuring protein concentration in egg white using the biuret method

Here is another series of examples in which adding specific details improves the title:

FAULTY: Thermoregulation in chicken embryos

FAULTY: Oxygen consumption of chicken eggs

EXPLANATION: More information is required to understand the context.

REVISION: Effect of temperature on oxygen consumption of 14-day old chicken embryos *or* Measuring oxygen consumption to understand how temperature is regulated in 14-day old chicken embryos

Here is another example in which the title is made more descriptive by removing unnecessary words and adding the specific variable that was manipulated:

FAULTY: Explanation of seed germination in *H. vulgare*

EXPLANATION: Avoid using "filler phrases" such as "explanation of," "analysis of," and "study of." Give the common name and the scientific name of the organism for the reader's benefit. Focus on the specific aspect of seed germination that you studied.

REVISION: Effect of gibberellic acid concentration on starch remaining in the endosperm of barley (*Hordeum vulgare*) seeds

Check Your Understanding

1. Skim a published journal article recommended by your instructor. Access "Instructions to Authors" on the journal's webpage. Read the detailed instructions that authors must follow to have their work considered for publication. Write a brief reflection on these instructions.

2. After writing your lab report (or a certain section), compare the writing style and content with that in the corresponding section(s) of a journal article recommended by your instructor. Make a checklist for each section, to remind yourself of things to do and not do.

Go to **macmillanlearning.com/knisely6e**
and select **"Student Site"** to access
samples, template files, and tutorial videos

## Chapter 6

Data Analysis Using Statistics

Claire Kelling, *Pennsylvania State University Department of Statistics*

Objectives

6.1 Distinguish between quantitative and categorical variables

6.2 Learn how data exploration can identify gaps or outliers in the data

6.3 Understand how to interpret histograms, box plots, and scatterplots

6.4 Explain why hypotheses must be proposed before data are explored

6.5 Learn the four-step procedure for hypothesis testing

6.6 Understand when to use t-tests, chi-square tests, and correlations

6.7 Understand the advantages of using R for data analysis

6.8 Learn useful R code to adapt to other datasets

6.9 Understand some of the ethical concerns in statistical data analysis

6.1 Introduction

Data collected through carefully designed experiments allow biologists and scientists more generally to quantitatively explore biological phenomena. For example, biologists might be interested in the relationship between a certain drug and a health outcome, such as tumor size. Collecting data allows us to explore the effect of the drug on increasing or decreasing tumor size, for example, or to determine whether it has no effect.

Statistics allow biologists to draw conclusions about phenomena based on the dataset that was collected. Without statistics, biologists would only be able to describe what they see as an increase, decrease, or no change in tumor size, in our example above. With statistics, we are able to determine whether the difference between treatments is statistically significant or a result of random chance. The field of statistics allows data and hypotheses to be translated into actionable conclusions.

A large variety of statistical tests is used in many applications in biology. Statistics are used quite heavily in ecology, neuroscience, genomics, and psychology. Fairly simple statistical tests can be used to make important and powerful conclusions. For example, Barnett et al. (2005) used t-tests

117

and chi-square tests to explore the potential of regulatory T cells in treating ovarian cancer in humans. More complex statistical models incorporate additional dynamics, such as space or time, that may affect a study. For example, Liang et al. (2008) studied the spatial distribution of cancer in Minnesota while controlling for community level factors such as poverty and whether the case was in a metro area. Statistical models are useful in testing a wide range of hypotheses and research questions.

In this chapter, we outline some of the foundational statistical tests used in the biological sciences. We describe how a student could go about exploring and analyzing a dataset from an experiment in an introductory college biology lab. We describe the statistical questions we are interested in and how to interpret results to form conclusions. For statistical tests, we show how to conduct t-tests for one sample and two sample means of numeric variables and chi-square tests for categorical variables. We apply these methods to example datasets. Finally, we discuss issues of broader interest to scientists and statisticians, such as software programs for statistical analysis and ethical concerns.

6.2 Definitions, Symbols, and Equations

Some of the terms, symbols, and equations used in this chapter are provided in the following lists.

Experimental design

- **Population**: The entire set of possible cases of interest.
- **Sample**: A subset of the population for which we are able to collect data and from which we can draw conclusions about the population.
- **Replicates**: Repetitions of certain experimental conditions, often used to assess variability within these conditions.

Data types

- **Continuous:** A variable that takes on numeric values with an infinite number of possible values (e.g., weight, temperature, distance).
- **Categorical:** A variable that takes on discrete values (e.g., ethnicity, type of tree, high/low).
- **Ordinal:** A type of categorical data, where the levels are ordered (e.g., agree, neutral, or disagree).

Descriptive statistics

- **Sample size** (n): The number of samples that were observed or measured.

- **Outlier:** An observation that notably differs from the rest of the observed or measured data.
- **Sample mean**: If $x_1, \dots, x_n$ are the observed or measured data, then the sample mean, $\bar{x}$, is calculated as follows:

$$\bar{x} = \frac{1}{n} \sum_{i=1}^{n} x_i$$

- **Percentile**: The pth percentile is the value at which p percent of the data falls below this value. For example, if 150 cm is the 30th percentile in height, that means 30% of the height data falls below 150 cm.
- **Quartile**: Divides the data into four groups. The complete set of quartiles consists of the 25th, 50th, and 75th percentiles.
- **Median**: If our numeric observations $x_1, \dots, x_n$ are *ordered* from lowest to highest, the median is the value in the middle of the distribution, or $x_{n/2}$ when n is even or $x_{(n+1)/2}$ when n is odd. The median is another measurement of the center of the observations, but it is less sensitive to outliers than the mean. The median is the 50th percentile.
- **Range**: The maximum to minimum value for a given numeric variable.
- **Sample standard deviation**:

$$s = \sqrt{\frac{\sum_{i=1}^{n} (x_i - \bar{x})^2}{n - 1}}$$

where $x_1, \dots, x_n$ are the data, $\bar{x}$ is the sample mean, and n is the sample size. Used to measure variability within a dataset.
- **Standard error of the mean**:

$$SEM = \frac{s}{\sqrt{n}}$$

where s is the sample standard deviation. Used to show how far the sample mean is likely to lie from the true population mean.

Hypothesis testing

Scientists develop research questions that they hope to answer by conducting an experiment or an observational study. The experiment generates data that scientists must organize and interpret. Before data analysis, they propose statistical hypotheses called the null and alternative hypotheses.

- **Null/alternative hypotheses**: The **null hypothesis** is typically that all is equal, or there are no differences between groups or from a hypothesized value. The **alternative hypothesis** usually states that there are differences between groups or from a hypothesized value, and the observed difference is not due to chance but rather is statistically meaningful. For example, we could propose a null hypothesis that the mean heights of mature maple and oak trees are equal. The alternative hypothesis could be that the mean heights of maple and oak trees are not equal. Because the mean height of maple trees could be less than or greater than that of oak trees, this is an example of a two-sided test. A different alternative hypothesis, depending on the research question, is that the mean height of maple trees is greater than that of oak trees (or vice versa). Both of these alternative hypotheses are examples of one-sided tests. Statistical hypotheses are usually denoted H_0 for the null hypothesis and H_1 or H_a for the alternative hypothesis.

- **Confidence Interval**: A confidence interval is defined as a range of values around the sample statistic, where the range is partially determined by the confidence level, for example, a 99% confidence level. The true population parameter falls within this percentage of calculated confidence intervals, in this case, 99% of calculated confidence intervals, when assumptions of the test are valid.

- **Type I Error**: A type I error results when we reject the null hypothesis, but the null hypothesis is true; also called a false positive.

- **Type II Error**: A type II error results when we accept the null hypothesis, but the alternative hypothesis is true; also called a false negative.

6.3 An Experiment about Blackworms

To illustrate how experimental data might be analyzed statistically, we will use a dataset on blackworms collected by students in an organismal biology laboratory at Bucknell University in 2004. The purpose of this experiment was to study how various drugs affect the circulatory system in blackworms (*Lumbriculus variegatus*). These small worms have a transparent body wall, which makes it possible to count the waves of blood pulsation under a microscope without any invasive procedures. The pulsation rate (measured in beats/min) was measured for an individual blackworm in spring water. The worm was then placed in a beaker with a drug solution for 15 minutes. After 15 minutes, the pulsation rate was measured again. The percentage change in pulsation rate was calculated using this formula:

$$\frac{Rate\ after - Rate\ before}{Rate\ before} \times 100$$

Two drugs were tested on the blackworms: caffeine and alcohol, each at five levels. Alcohol was tested at concentrations of 0.5%, 1%, 2%, 3%, and 4% (parts per hundred). Caffeine was tested at concentrations of 5 mM, 10 mM, 15 mM, 20 mM, and 25 mM (millimolar or millimoles per liter). The difference in units is important to note because this will affect the ability to make comparisons.

In humans, caffeine-containing beverages can act as a stimulant, improve mood, and increase work capacity. Alcohol is primarily a central nervous system depressant. We used these known drug effects in humans to propose experimental hypotheses about blackworms. Specifically, we hypothesized that caffeine would act as a stimulant and alcohol as a depressant in blackworms. We predicted that if the hypotheses were correct, blackworms exposed to caffeine would have an increased pulsation rate compared with the control; those exposed to alcohol would have a decreased pulsation rate.

We will use the blackworm dataset throughout this chapter to show how to apply exploratory tools and statistical tests. See the link at the end of this chapter to access the original Excel file and R code. Other example datasets were constructed for illustrative purposes and do not represent actual measured data.

In this chapter, we will analyze our data using R. R is a free, open source software program that is widely used by statisticians and scientists (R Core Team 2017). There is a vast community of R users around the world who collaborate online, making R one of the fastest growing programming languages (Ihaka 2017). In fact, the R programming community includes more than 12,340 contributors from academia, private industry, government, and other fields (Korkmaz 2018). For the examples in this chapter, we include code snippets inline, so the reader may repurpose similar analysis and code in their own work. In most cases, we include the code immediately followed by the output. We rely on the *tidyverse* package for the majority of our programming, which allows for us to easily import, clean, and visualize data (Wickham et al. 2019).

We will introduce some details of useful conventions for coding in R. However, for a full introduction, we recommend the *R Cookbook* textbook that provides many examples from R basics to creating graphics and running statistical tests and models (Teetor 2011). We provide the formulas to calculate test statistics for the t-test and chi-square test by hand, but in the real world, scientists and statisticians will use statistical software to compute these statistics. In each case, we will illustrate how to interpret the output from R.

6.4 Data Exploration

When we analyze a dataset, it is important to have research questions and hypotheses prepared *before* exploring the data. If we analyze data before developing research questions, we may observe correlations that are not associated with causal mechanisms. A classic example involves the observation that ice cream sales, drowning incidents, and forest fires all tend to increase in the summer months (Feenstra and Hoffman 2015). Just because there may be a correlation between these three events, it does not mean that one event causes either of the others. On the other hand, when research questions guide how a study is carried out, scientists consider many possible variables that could affect the response. In this example, the causal mechanism is likely the heat, which causes more people to buy ice cream, more people to swim and possibly drown, and drier conditions, which lead to forest fires.

Data exploration is a useful tool to gain a better understanding of your datasets. This process can help reveal any flaws in the data collection process, which may not be visible on examination of the raw data. For example, missing data or data entry errors often become easily visible as a result of thorough data exploration.

Data exploration also involves data visualization to spot trends and outliers more easily. Useful plotting tools include boxplots, histograms, and scatterplots. Subsequently, data exploration can help inform what sorts of statistical tests or models should be considered. Data exploration is often not included in the results sections of published studies because the point is not to conduct statistical tests, but certain plots may be appropriate for inclusion in reports, depending on the application.

For the blackworm dataset, our data were collected in Excel. We read the data from Excel into R using the following code, where the two sheets in Excel are labeled "Alcohol" and "Caffeine." We converted the data to a data frame to assist with analysis. A **data frame** is a flexible way to store data in R, similar to a matrix in other programming languages. It allows for easy manipulation from the *tidyverse* R package.

```
alcohol_data <- read_xlsx("Blackworms_Claire.xlsx",
    sheet = "Alcohol") %>% as.data.frame()
caffeine_data <- read_xlsx("Blackworms_Claire.
    xlsx", sheet = "Caffeine")%>% as.data.frame()
```

Many tools for data exploration are available in R. We will use the *summary()* function in R to illustrate some of the information we can obtain about the dataset. As shown in the following output, three variables in the blackworm alcohol dataset are stored in a data frame. The first is the treatment type, namely, alcohol. In our example, the alcohol and caffeine datasets are stored in separate data frames because they have different units of measurement. The second variable in each data frame is the concentration, measured as a percentage (alcohol) or in millimoles per liter (caffeine). The third variable is the pulsation change corresponding to the concentration and treatment for that sample. The output for the alcohol dataset shows three columns (treatment, alcohol concentration, and pulsation change) and indicates that 108 samples were collected in the course of the experiment. We note that there are many replicates under the same experimental conditions. For example, there are 16 replicates for the 0.5% alcohol treatment.

Many types of variables are stored in R. The most common are character, numeric, integer, factor, and logical variables. Character variables are often words or combinations of words, such as names. Numeric and integer variables are self-explanatory. In our example, the concentration of alcohol and the pulsation change are both numeric variables. Both could be rounded and converted to integer data types, although this is not recommended unless the data actually only take integer values. Factors are categorical variables with given levels. If the dataset included both alcohol and caffeine, the treatment variable could be considered a factor variable with two levels. Logical variables take the value *true* or *false*. The *summary()* function gives details on the variables that are included in the dataset and some preliminary distribution information for quantitative variables, such as quartiles, the minimum, the maximum, and the mean. The *summary()* function also gives some details on missing data. In the third column, the "1" next to "NA's" tells us that there is one missing value for the pulsation_change variable.

```
# Exploratory Data Analysis
## How many rows and columns
dim(alcohol_data)
[1] 108    3
## Summary of dataset
summary(alcohol_data)
```

treatment	concentration_perc		pulsation_change		
Alcohol:108	Min. :	0.500	Min.	:	-55.00
	1st Qu. :	1.000	1st Qu.	:	-16.82
	Median :	2.000	Median	:	12.80
	Mean :	2.102	Mean	:	18.41
	3rd Qu. :	3.000	3rd Qu.	:	49.00
	Max. :	4.000	Max.	:	153.73
			NA's	:	1

The *filter()* function is a useful function in R to remove or isolate rows in the dataset based on certain criteria. The *is.na() function* tests for whether a certain variable contains missing values, which are coded in R as "NA." The following code allows us to remove rows from both the alcohol and caffeine datasets in which values for the pulsation_change variable are missing. These values were probably omitted because the student was unable to measure the pulsation rate under both conditions (spring water and after 15 minutes in the drug solution).

```
# Removing missing data
alcohol_data <- alcohol_data %>% filter(!is.
     na(pulsation_change))
caffeine_data <- caffeine_data %>% filter(!is.
     na(pulsation_change))
```

To access a variable in a data frame, the $ operator can be used along with the variable name, as shown in the following code snippet. The *class()* function determines the type of variable in a data frame. In our example, we show that the treatment variable is a character variable when it is read from the original dataset. If we want to convert the treatment variable to a factor variable for analysis, we can reassign the variable by using the *as.factor()* function. There are equivalent functions to convert data to other variable types: *as.numeric(), as.character()*, etc.

```
## Determine type of variables
class(alcohol_data$treatment)
[1] "character"
## Changing variable type
alcohol_data$treatment <- as.factor(alcohol_
      data$treatment)
```

To explore the dataset quickly, we can use the *head()* function to show the first few rows in the dataset and all columns. The second argument, n, gives the number of rows to display. In the first few rows of the black-worm dataset with the alcohol treatment, we can quickly see the repetition of the alcohol treatment, the use of replicates for the 0.5% concentration, and the measurements of pulsation change. In more complex datasets that we did not collect, perhaps with dozens of variables, a quick glance at the data would allow us to see what general form the data take.

```
## Quick view of dataset
head(alcohol_data, n = 5)
```

	treatment	concentration_perc	pulsation_change
1	Alcohol	0.5	-30.600
2	Alcohol	0.5	-44.000
3	Alcohol	0.5	17.850
4	Alcohol	0.5	-3.125
5	Alcohol	0.5	28.000

We note that samples for both the alcohol and caffeine treatments are collected with replicates at distinct concentration levels. For example, replicates are collected at concentrations of 0.5%, 1%, 2%, 3%, and 4% alcohol. Although this could be seen as a categorical variable because the levels are distinct, we treat this variable as a continuous numeric variable.

Histograms

Histograms are useful tools for exploring the distribution of a single variable, for example, change in pulsation rate. A **histogram** shows the frequency of observations along a range of values. The appearance of a histogram is dependent on the number of breaks, or bins, provided by the user. Figure 6.1 shows histograms with 10 breaks (A) and 50 breaks (B). More samples are included in each frequency bar when there are fewer breaks or when the bin width is larger.

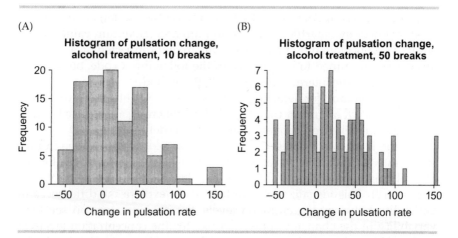

Figure 6.1 Histograms display the frequency of each change in pulsation rate for 107 blackworms treated with alcohol. (A) With 10 breaks (or bins), the bars are wider and each bar represents more observations. (B) The same data displayed with 50 breaks.

```
#Histogram of data
hist(alcohol_data$pulsation_change, breaks = 10,
     main = "Histogram of Pulsation Change,
      \nAlcohol Treatment, 10 breaks",
   xlab = "Change in pulsation rate")
hist(alcohol_data$pulsation_change, breaks = 50,
     main = "Histogram of Pulsation Change,
      \nAlcohol Treatment, 50 breaks",
   xlab = "Change in pulsation rate")
```

Both histograms let us see that we have both positive and negative values for the change in pulsation rate, with a few values as high as 150, which may be considered outliers. We see that most of the data values lie between -25 and 50 beats/min. In general, we should choose a number of breaks that allows us to summarize the distribution, without having a bar for each observation. In this case, the histogram with 10 breaks looks coarser and shows the overall trends more clearly than the histogram with 50 breaks, which is more granular and would show smaller differences between observations.

Boxplots

Boxplots are another simple way to visualize the distribution of a variable. The middle line in the center of the "box" in a boxplot is the median, or the second quartile. The box edges at either end of the box represent the first and third quartiles. The lines at either end of the boxplot, or the "whiskers," connect the quartiles to the minimum and maximum values. If we observe points plotted by themselves that are not connected by a whisker, these are determined to be outliers. In the following R code, outliers are determined by the interquartile range (IQR) method.

The **IQR method** for detecting outliers defines an outlier as a value less than $Q_1 - 1.5IQR$ or greater than $Q_3 + 1.5IQR$, where Q_1 and Q_3 are the first and third quartiles, respectively, and the IQR is defined as $Q_3 - Q_1$.

For the blackworm dataset, we create a boxplot of the change in pulsation rate for each concentration level, measured as percentages for the alcohol treatment and in millimoles per liter for the caffeine treatment (Figure 6.2). This visualization can help us to explore the differences in variability between concentration levels. For example, we can see high variability in the change in pulsation rate for the concentration levels 15 mM and 25 mM for the caffeine dataset. We can also see variability in the change in pulsation rate at a concentration of 20 mM, but the minimum and maximum are determined to be outliers by the IQR method.

```
#Boxplots of each concentration
alcohol_plot <- ggplot(alcohol_data, aes(x=as.
      factor(concentration_perc), y=pulsation_
      change)) +
   geom_boxplot()+
   stat_summary(fun=mean, geom="line",
      aes(group=1), col = "blue")   +
   stat_summary(fun=mean, geom="point", col =
      "blue")+
   labs(x= "Concentration (%)", y = "Change
      in pulsation rate", title = "Treatment:
      Alcohol")
caffeine_plot <- ggplot(caffeine_data, aes(x=as.
      factor(concentration_mm), y=pulsation_
      change)) +
   geom_boxplot()+
   stat_summary(fun=mean, geom="line",
      aes(group=1), col = "blue")   +
   stat_summary(fun=mean, geom="point", col =
      "blue")+
   labs(x= "Concentration (mM)", y = "Change
      in pulsation rate", title = "Treatment:
      Caffeine")

#Plot alcohol and caffeine together
grid.arrange(alcohol_plot, caffeine_plot, nrow=1)
```

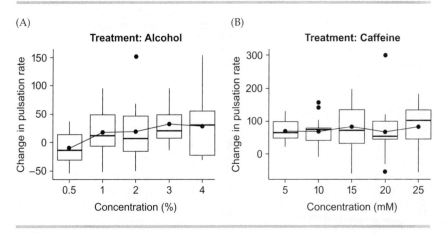

Figure 6.2 Boxplots display distribution of pulsation rates by concentration level. (A) Effect of different concentrations of alcohol as percentage. (B) Effect of different concentrations of caffeine expressed as millimoles per liter (mM).

Scatterplots

When we are interested in exploring the relationship between two variables, scatterplots provide a method of plotting one variable against another. Each point in a scatterplot represents an observation or a sample. The scatterplot for the alcohol treatment (Figure 6.3) shows the concentration plotted on the x-axis and the change in pulsation rate on the y-axis. Each point represents a measurement taken for an individual blackworm. We see that the concentration is taken at fixed values and there is a wide range in the change in pulsation rate for each concentration level. Neither pattern is so apparent from looking at the numbers in the spreadsheet. This illustrates the importance of using exploratory tools before conducting formal hypothesis testing because we can see that there are replicates taken at each concentration, and this might not have been clear otherwise.

```
#Scatterplot of data
plot(alcohol_data$concentration_perc, alcohol_
        data$pulsation_change,
    xlab = "Concentration (%)", ylab = "Change in
        pulsation rate",
    main = "Scatterplot of Concentration vs Change
        in Pulsation, \nAlcohol Treatment")
```

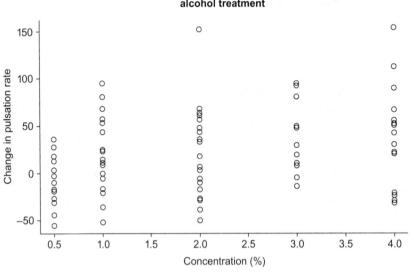

Figure 6.3   Scatterplots show the relationship between two variables, in this case, between alcohol concentration and change in pulsation rate for blackworms.

6.5 Hypothesis Testing

After exploring our data, we can conduct hypothesis tests to answer research questions about the data. The research question and data format will determine the exact statistical test to use. For example, if data are numeric and we are interested in the mean of the data, we will typically use either a t-test or a z-test. If the data are categorical and our research question concerns the distribution of the data with respect to the categories, we would use a chi-square test.

Test statistic

A **test statistic** is a numeric summary of your data, which is used to test a hypothesis. We see test statistics in all of the following tests, including the t-test, chi-square test, and z-test. The test statistic is used, sometimes along with additional information about the data, to reject or fail to reject the null hypothesis. Just because we do not reject a null hypothesis, it does not necessarily mean that the null hypothesis is true. It simply means we do not have enough evidence in our data to reject the null hypothesis. The name of the test is the name of the distribution that is used in the hypothesis testing procedure (either t, z, or chi-square distribution). Generally, a test statistic is calculated as follows:

$$\frac{sample\ statistic - hypothesized\ value}{sample\ standard\ error}$$

The sample statistic and the sample standard error are calculated using the measured or observed data. The hypothesized value is typically motivated by the research questions and is therefore often based on prior scientific knowledge or belief. For example, if we are trying to determine whether alcohol has an effect on the pulsation rate, we would test to see whether the sample statistic, in this case the sample mean of the change in pulsation rates, is different from 0. More specifically, if we wanted to determine whether alcohol acts as a stimulant, we would determine whether the mean change in pulsation rate is greater than 0. If the change in pulsation rate is equal to 0, this would indicate that alcohol has no effect on pulsation rate.

A t-test for one sample mean

The purpose of a t-test is to statistically compare a sample mean to a hypothesized value. The t-distribution is used when the population is normally distributed but the population standard deviation is unknown or when the population is not known to be normally distributed but the sample size is large ($n \geq 30$). If the population follows a normal distribution and the population standard deviation is known, then a z-distribution is used.

The first step in the four-step procedure for a t-test is to **establish the hypothesis** based on the research question. With a t-test, we can test whether the population mean, μ, is equal to a hypothesized value, μ_0. The null hypothesis (H_0) and the alternative hypothesis (H_a or H_1) for each research question are shown in the following table:

Research Question	H_0: Null Hypothesis	H_1: Alternative Hypothesis
Is the mean equal to μ_0?	$\mu = \mu_0$	$\mu \neq \mu_0$
Is the mean greater than μ_0?	$\mu = \mu_0$	$\mu > \mu_0$
Is the mean less than μ_0?	$\mu = \mu_0$	$\mu < \mu_0$

Second, calculate the test statistic using the data. The test statistic t is calculated using the sample mean, $\bar{x}$, the hypothesized population mean, μ_0, and the sample standard deviation, s.

$$t = \frac{\bar{x} - \mu_0}{\frac{s}{\sqrt{n}}}$$

Third, find the p-value based on the test statistic. The p-value is the probability of getting the results we observed, which are summarized by the test statistic, under the null hypothesis. A small p-value indicates that there is a low probability that the results we observed would occur under the null hypothesis, or that there is evidence against the null hypothesis. A large p-value indicates that our results were reasonable under the null hypothesis. The p-value can be determined using R or a t-table. You will need to know the *degrees of freedom* associated with the t-test, which is calculated as $df = n - 1$.

Finally, make a decision and state the conclusion based on your data. We choose a given significance level, alpha (α), where if the p-value is less than α, we reject the null hypothesis. Otherwise, we fail to reject the null hypothesis. Typical values of α are 0.05 or 0.01. In our case, we choose α to be 0.01 because we would like strong evidence to reject the null hypothesis to avoid a type I error (a false positive).

There are a couple of additional considerations when stating your conclusion(s). First, it is important to not only state what the p-value shows but also to include any caveats of your experiment or data collection process. Caveats could include limitations of the sampling design and the acknowledgment that it is difficult to make causal conclusions based on observational data. In addition to acknowledging potential caveats, it is also important to not rely exclusively on p-values, as outlined later in Section 6.7.

Now we will apply the t-test to the blackworm dataset, following the procedure we just described.

1. **Establish the hypothesis:** For the alcohol dataset, we ask whether the mean change in pulsation rate is equal to 0, resulting in the following hypotheses: $H_0: \mu = 0, H_1: \mu \neq 0$.

2. **Calculate the test statistic:** This step is easily done in R through the *t.test()* function as shown in the following code. The first argument, *x*, is the data that is used to calculate the test statistic. The second argument states the alternative hypothesis. In our case, we have a two-sided alternative hypothesis; other possible arguments for alternative include "less" or "greater." Lastly, the argument *mu* states the hypothesized value that we are testing against, or 0 in this case.

```
t.test(x=alcohol_data$pulsation_change,
       alternative = "two.sided", mu = 0)
```

The output for this code is as follows:

```
        One Sample t-test
data:   alcohol_data$pulsation_change
t = 4.2377, df = 106, p-value = 4.834e-05
alternative hypothesis: true mean is not equal to 0
```

```
95 percent confidence interval:
      9.79581  27.02035
 sample estimates:
 mean of x
     18.40808
```

This output gives all of the information needed to make a conclusion using our data in combination with our hypotheses. We note that in the second line of output, in the solid box, we see the test statistic, degrees of freedom, and p-value. The last line of output, in the dashed box, also includes the sample mean.

3. **Find the p-value:** Based on the output in step 2, we find that the p-value is very small, shown to be 4.834×10^{-5}.

4. **Make a decision and state the conclusion:** Because the p-value is less than $\alpha = 0.01$, we are able to reject our null hypothesis that the mean of the change in pulsation rate is 0 for the alcohol treatment. Therefore, we can say that through these results, we see that alcohol does have an effect on the pulsation rate. Possible caveats for this experiment are that the experiment was conducted on lab-raised blackworms and the results may not apply to worms in nature. We also did not control for size, age, or health of the worms. We also cannot say whether these results apply to other animals, such as humans.

To illustrate how p-values can be misused or "hacked," let's assume, for this example, that we are intent on proving that alcohol has no effect on pulsation rate because it will get our paper published. We could keep testing additional combinations of worms until we calculate a test statistic that gives us a p-value that is greater than α. This an example of p-hacking and is unethical.

As noted previously, if the population standard deviation is unknown, the t distribution should be used. However, if the population can be assumed to be normally distributed and the population standard deviation is known, the z distribution can be used, which does not rely on degrees of freedom. The assumption that the population is normally distributed is usually based on prior knowledge of the outcome of interest. The test statistic for the z distribution is very similar, where σ is the known population standard deviation.

$$z = \frac{\bar{x} - \mu_0}{\frac{\sigma}{\sqrt{n}}}$$

The R function for the sample mean z-test is *z.test()* in the BSDA R package and has the same inputs as *t.test()*.

A t-test for two sample means

So far in this section, we have shown an example for testing one sample mean against a hypothesized value. However, we note that the t-test is also often used to test *independent means* when we are interested in the difference between two datasets. For example, in the blackworm data, we might be interested in the difference in the mean change in pulsation rate between the 1% and 2% alcohol solutions.

When testing the difference between two means, the procedure outlined for one sample mean applies, but the test statistic and hypotheses are slightly different. Here is how the procedure for one sample mean would be adapted to comparing two means.

1. **Establish the hypothesis:** The null hypothesis is that the two means are equal (H_0: $\mu_1 = \mu_2$), while the alternative hypothesis can once again be one- or two-sided (H_1: $\mu_1 \neq \mu_2$, $\mu_1 > \mu_2$, or $\mu_1 < \mu_2$). These hypotheses also equivalently determine whether the difference between the means is 0 (H_0: $\mu_1 - \mu_2 = 0$).

2. **Calculate the test statistic:** Through the parametrization of the hypothesis shown in step 1, we see why the test statistic relies on the difference between the sample means, or $\bar{x}_1 - \bar{x}_2$, and the associated standard deviation. The degrees of freedom associated with the test, when we do not assume that there is equal variance between the two samples, is calculated within the R function.

$$t = \frac{\bar{x}_1 - \bar{x}_2}{\sqrt{\dfrac{s_1^2}{n_1} + \dfrac{s_2^2}{n_2}}}$$

To test the hypothesis if the mean change in pulsation rate between the 1% and 2% alcohol solutions is the same (H_0: $\mu_1 = \mu_2$) vs (H_1: $\mu_1 \neq \mu_2$), we would run the following code in R, where x is the 1% solution and y is the 2% solution.

```
#subset the data
one_perc_data <- alcohol_data %>%
      filter(concentration_perc == 1)
two_perc_data <- alcohol_data %>%
      filter(concentration_perc == 2)
#run the t-test
t.test(x=one_perc_data$pulsation_change, y = two_
      perc_data$pulsation_change, alternative =
      "two.sided", mu = 0)
```

```
Welch Two Sample t-test
data:   one_perc_data$pulsation_change and two_
        perc_data$pulsation_change
t = -0.10527, df = 46.181, p-value = 0.9166
alternative hypothesis: true difference in means
        is not equal to 0
95 percent confidence interval:
        -27.29085   24.57789
sample estimates:
mean of x mean of y
      17.91852    19.27500
```

3. **Find the p-value:** As shown in the output of this code, the test statistic is -0.10527 and the associated p-value is 0.9166.

4. **Make a decision and state the conclusion:** Because the p-value is greater than $\alpha = 0.01$, we fail to reject the null hypothesis that the mean change in pulsation range is equal between the 1% and 2% levels of alcohol.

Chi-square test for one variable

Whereas t-tests are often used to test a sample mean or two sample means for continuous numeric data, chi-square tests are used to test categorical variables. We often encounter categorical variables—such as blood type, flower color, and wing shape—when working out genetics crosses, which illustrates why chi-square tests are frequently used in genetics.

Categorical data are often presented in a frequency table, where the number in each cell of the table represents the number of individuals or samples that fall into that category. Table 6.1 is an example of a one-way frequency table, with one categorical variable, where we show the frequency of different tree types in a sample. To collect these data, we could walk around part of a property and count the trees of each type. We would like to know whether this sample follows our hypothesis of what we believe the tree distribution is like on the entire property.

TABLE 6.1   Number of trees counted on a walk around one part of a property. This is an example of a one-way frequency table.

Tree Type	Frequency
Oak	53
Maple	74
Pine	3
Total	130

TABLE 6.2	Expected number of trees based on our hypothesis
Tree Type	**Expected Frequency**
Oak	130*0.4 = 52
Maple	130*0.5 = 65
Pine	130*0.1 = 13
Total	130

Next, we outline the same four-step procedure for the chi-square test that we outlined previously for the t-test. We will discuss the application in each step.

1. **First, to establish the hypothesis,** we establish the proportions that we believe are true for all of the trees on the property. For our application, we believe that 40% are oak, 50% are maple, and 10% are pine. So, our null hypothesis is as follows: H_0: $p_{(oak)} = 0.4$, $p_{(maple)} = 0.5$, $p_{(pine)} = 0.1$. Our alternative hypothesis is that at least one of these probabilities is not correct.

2. **To calculate the test statistic for a chi-square test**, we need to compare what we observed to what we expected to observe, based on our hypothesis. To find the expected number of counts for each category, or level of the categorical variable, we multiply the total number of observed trees (130) by the probability for each category under the null hypothesis. The expected frequencies are shown in Table 6.2.

After calculating the expected frequencies for each category, we can calculate the chi-square test statistic as follows, where we have c categories.

$$x^2 = \sum_{i=1}^{c} \frac{(Observed_i - Expected_i)^2}{Expected_i}$$

In our case, the chi-square test statistic is calculated as follows:

$$x^2 = \frac{(53-52)^2}{52} + \frac{(74-65)^2}{65} + \frac{(3-13)^2}{13} \approx 8.96$$

We find the degrees of freedom for our application, where $df = c - 1 = 3 - 1 = 2$.

3. **To find the p-value**, we can use a chi-square table or the test in R. The R code and the output for this test follow. The first argument in the R function is the vector of observations. The second argument is the vector of hypothesized probabilities. In the boxed section of the output, we find the same test statistic that we calculated above, as well as the p-value associated with the test statistic, 0.01135.

```
chisq.test(x = c(53, 74, 3),
           p = c(0.4, 0.5, 0.1))
```

Chi-squared test for given probabilities

data: c(53, 74, 3)

X-squared = 8.9577, df = 2, p-value = 0.01135

4. To **make a decision and state the conclusion,** first we compare the p-value to $\alpha = 0.01$. We find that the p-value is greater than α, although close, so we fail to reject the null hypothesis. We conclude that there was not enough evidence to reject our null hypothesis of proportions of oak, maple, and pine trees to be 40%, 50%, and 10%, respectively. However, we should also acknowledge the potential caveats in our sampling methods that might have affected our analysis. For example, we might have surveyed only one small part of the property instead of multiple spots throughout the property.

Chi-square test for two variables

Chi-square tests can also be useful to test the relationship between two categorical variables. Let's say we want to analyze the frequency data for two types of animals, snowshoe hares and Canada lynxes, to determine whether there might be a relationship between them. For our study, we analyze 355 images from motion-activated wildlife cameras set up in a large forest in British Columbia and list the observed frequencies in a two-way frequency table. Each cell in the table is the sum of the number of times we observed only hares, only lynxes, both, or neither in our sample. Table 6.3 shows the observed frequencies of hares and lynxes for our example.

To quantify the distribution of these two types of animals, we follow the analogous procedure described previously, but for the chi-square test of independence between two variables.

1. **Establish the hypothesis:** In this case, we are testing the null hypothesis that there is no significant difference in the distribution of these two animals, or the distribution is random. The

TABLE 6.3 Example data for the observed frequencies of snowshoe hares and lynxes. This is an example of a two-way frequency table.

	Canada lynx		
Snowshoe hare	Present	Absent	Row Total
Present	212	32	244
Absent	50	61	111
Column Total	262	93	355

TABLE 6.4 Example data for the expected frequencies of snowshoe hares and lynxes. This is an example of a two-way frequency table.

Snowshoe hare	Canada lynx		Row Total
	Present	Absent	
Present	(262*244)/355 ≈ 180.1	(93*244)/355 ≈ 63.9	244
Absent	(262*111)/355 ≈ 81.9	(93*111)/355 ≈ 29.1	111
Column Total	262	93	355

alternative hypothesis is that there is an association between the two species.

2. **Calculate the test statistic**: As before, we need to compare what we observed to what we expected to observe, based on our hypothesis. To find the expected number of counts for each cell, we multiply the row and column totals and divide by the total number of samples, 355. The expected frequencies are shown in Table 6.4.

After calculating the expected frequencies for each category, we can calculate the chi-square test statistic as follows, where we have c combinations of categories, or four in our example. This is the same as the number of cells in the two-way table, not including totals.

$$x^2 = \sum_{i=1}^{c} \frac{(Observed_i - Expected_i)^2}{Expected_i}$$

In our case, the chi-square test statistic is as follows:

$$x^2 = \frac{(212-180.1)^2}{180.1} + \frac{(32-63.9)^2}{63.9} + \frac{(50-81.9)^2}{81.9} + \frac{(61-29.1)^2}{29.1} \approx 68.97$$

We find the degrees of freedom for our application, where $df = (number\ of\ rows - 1)(number\ of\ columns - 1)$ to be $(2-1)(2-1) = 1$.

3. **Find the p-value**: Again, we can use a chi-square table or the test in R. The R code and the output for this test follow. The first argument in the R function is the matrix of observed values and the second is to indicate that we do not want to implement a continuity correction. In the output, which is shown in the box, we find a test statistic similar to the one we calculated earlier, as well as the p-value associated with the test statistic. The test statistic is slightly different as a result of rounding error in our table of expected counts, which is another reason why using R is advantageous.

```
chisq.test(x = matrix(c(212, 50, 32, 61), nrow = 2),
      correct = FALSE)
```

```
Pearson's Chi-squared test
data:   matrix(c(212, 50, 32, 61), nrow = 2)
X-squared = 69.079, df = 1, p-value < 2.2e-16
```

4. **Make a decision and state the conclusion:** We compare the p-value to $\alpha = 0.01$. The p-value is less than α, so we reject the null hypothesis. This conclusion means that there is a relationship between lynxes and snowshoe hares based on our sample. The wildlife camera images show that lynxes and snowshoe hares are often observed in the same locations. Therefore, we see a positive association between the two animals, which is common for predator-prey relationships. For caveats of our experiment, we must acknowledge that we only surveyed one forest in British Columbia, so this conclusion may not be valid in other parts of Canada or in the United States.

Confidence intervals

The hypothesis tests described in the previous section allow us to determine whether groups are equal or different. A confidence interval gives an idea of where we might expect our true, or population value, such as the population mean, to lie. Hypothesis testing and calculating confidence intervals are often done individually, but it may also be helpful to report a confidence interval when conducting hypothesis tests. A **confidence interval** is defined as a range of values around the sample statistic, where the range is partially determined by the confidence level, for example, a 99% confidence level. The true population parameter falls within this percentage of calculated confidence intervals, in this case, 99% of calculated confidence intervals, when assumptions of the test are valid. A confidence interval generally takes the following form, where the sample statistic is often the sample mean and the multiplier is used from the distribution of interest, such as the t distribution.

$$sample\ statistic \pm (multiplier \times standard\ error)$$

In our blackworms example, we may want to create a confidence interval around the mean of the pulsation change for all concentration levels. The sample statistic is then the sample mean, $\bar{x}$, and we use the t distribution if we assume that the standard error of the population is unknown to find the multiplier, t^*, for our level of confidence and the number of degrees of freedom. We will calculate confidence intervals for the blackworm dataset in the next section.

$$\bar{x} \pm \left(t^* \times \frac{s}{\sqrt{n}} \right)$$

6.6 Is There a Relationship between Alcohol or Caffeine Levels and Circulation in Blackworms?

To determine whether a relationship between alcohol or caffeine and circulation in blackworms exists, we can conduct a t-test based on the mean pulsation change. If the lower bound of the confidence interval associated with this t-test is greater than 0, we can be reasonably confident that the drug acts as a stimulant based on our experimental results. If the upper bound is less than 0, we can be reasonably confident that it acts as a depressant. As we calculated earlier, the sample mean for the change in pulsation rate in the alcohol treatment group is approximately 18.41 beats/min. The t-test that was conducted earlier actually also includes the 95% confidence interval for the sample mean under this number of degrees of freedom. If we wanted to calculate a confidence interval at a different level of confidence, such as 99%, we can add the additional *conf.level* argument to the *t.test()* function as follows:

```
t.test(x=alcohol_data$pulsation_change, alternative
    = "two.sided", mu = 0, conf.level = 0.99)
```

The output for this code is as follows:

```
One Sample t-test
data:   alcohol_data$pulsation_change
t = 4.2377, df = 106, p-value = 4.834e-05
alternative hypothesis: true mean is not equal to 0
99 percent confidence interval:
        7.013907 29.802251
sample estimates:
mean of x
        18.40808
```

Therefore, in our case, the 99% confidence interval for the sample mean is (7.014, 29.802). This interval indicates we can be 99% confident that the true mean pulsation rate for worms treated with alcohol lies between 7.014 and 29.802 beats/min. Because this confidence interval does not include 0, it is likely that alcohol increases the pulsation rate and therefore acts as a stimulant, based on the results from our experiment.

We can form a similar 99% confidence interval for the mean rate of change in pulsation for caffeine, as shown in the following code. The interval in the box also illustrates that it is likely that the true mean is greater than 0, and therefore, caffeine also increases the pulsation rate.

```
t.test(x=caffeine_data$pulsation_change,
        alternative = "two.sided", mu = 0, conf.level
        = 0.99)
```

```
One Sample t-test
data:   caffeine_data$pulsation_change
t = 11.963, df = 91, p-value < 2.2e-16
alternative hypothesis: true mean is not equal to 0
99 percent confidence interval:
      57.32594  89.65010
sample estimates:
mean of x
      73.48802
```

We can formally test the hypothesis that the mean is greater than 0 using a one-sided t-test where the alternative hypothesis is "greater" in the *t.test()* function, as shown in the following code for both alcohol and caffeine. We see that the p-value is very small for both alcohol and caffeine, and therefore, we reject the null hypothesis that the mean is equal to 0. Therefore, we can conclude that the mean change in pulsation rate is greater than 0 for both caffeine and alcohol. In other words, based on our sample, we can conclude that caffeine and alcohol are both stimulants at the concentrations tested.

```
t.test(x=alcohol_data$pulsation_change,
       alternative = "greater", mu = 0, conf.level
       = 0.99)
```

```
One Sample t-test
data:   alcohol_data$pulsation_change
t = 4.2377, df = 106, p-value = 2.417e-05
alternative hypothesis: true mean is greater than 0
99 percent confidence interval:
      8.147525       Inf
sample estimates:
mean of x
      18.40808
```

```
t.test(x=caffeine_data$pulsation_change,
       alternative = "greater", mu = 0, conf.level
       = 0.99)
```

```
One Sample t-test
data:   caffeine_data$pulsation_change
t = 11.963, df = 91, p-value < 2.2e-16
alternative hypothesis: true mean is greater than 0
99 percent confidence interval:
      58.94104       Inf
sample estimates:
mean of x
      73.48802
```

Correlation

To further investigate the relationship between alcohol and caffeine on circulation in blackworms, we can measure the correlation between the concentration of the treatment and the change in pulsation rate. Correlation measures the direction (positive/negative) and strength (weak/strong) of the linear relationship between two numeric variables. The correlation is bounded between -1 and 1. Negative values indicate a negative relationship and positive values indicate a positive relationship. Values of the correlation closer to -1 or 1 indicate a stronger relationship. This is not a formal hypothesis test, but rather serves as an exploratory tool to quantify the relationship between two variables. However, because correlation describes the strength of the linear relationship between two variables, it can be used to motivate the need, or lack thereof, for linear regression as a subsequent step in the analysis.

In the case of the blackworm dataset, we are interested in quantifying the relationship between alcohol concentration and the change in the pulsation rate. In the following R function and output, we find that the correlation between alcohol concentration and change in pulsation rate is approximately 0.23. This indicates a very weak positive relationship. For caffeine, we see a correlation of approximately 0.07, which shows an even weaker relationship. By default, R calculates the Pearson correlation coefficient, which measures the linear relationship between two variables.

```
cor(alcohol_data$concentration_perc, alcohol_
    data$pulsation_change)
```
0.2322783

```
cor(caffeine_data$concentration_mm, caffeine_
    data$pulsation_change)
```
0.06658812

When we add the linear regression line to the scatterplot shown in Figure 6.3, we visually confirm our finding of a weak positive relationship between alcohol concentration and the change in pulsation rate (Figure 6.4A). An even weaker relationship between caffeine concentration and pulsation rate is shown in Figure 6.4B.

Earlier in this section, we used t-tests to conclude that the mean change in pulsation rate is positive for the alcohol and caffeine treatments. The weak correlation allows us to conclude additionally that the change in pulsation rate does not increase with higher concentrations of alcohol or caffeine for the range of concentrations tested.

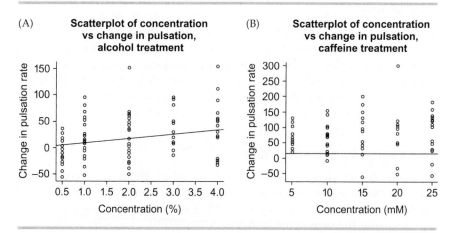

Figure 6.4 Scatterplots with the linear regression line added for all observations of concentration and change in pulsation rate. (A) Line of best fit for alcohol. (B) Line of best fit for caffeine.

6.7 Closing Thoughts

Technology

Many useful tools for conducting statistical analysis are available. As mentioned earlier, R is becoming the language of choice not only for statisticians but also for other scientists who are analyzing data. We use R for this chapter because of its flexibility and increasing popularity. Microsoft Excel is also a useful tool for conducting preliminary analyses and collecting and storing data. When conducting biological experiments, we will often collect and store the data in an Excel file because its data entry interface is easy to use. Excel can also be useful when conducting preliminary visualization exercises on data. We suggest that once data are collected, most data visualization and all data analysis should be conducted in R because of its flexibility and statistical rigor. Two other statistical software programs are JMP and Minitab. These are described as "point and click" software because you don't have to write code in these languages, but rather click on pre-programmed statistical tests. Although these programs provide useful tools for beginners, they are often not intuitive, and even the student editions are relatively expensive.

Although it is also valuable to know exactly how test statistics are calculated, as outlined in this chapter, we suggest using built-in R functions to calculate the test statistic and associated p-values, rather than calculating these values by hand. When calculating values by hand, it is easy to introduce simple errors along the way that can change your conclusion.

Rounding error can also have an important influence. In a typical introductory statistics class, you may find yourself using "z tables" to find the z-score test statistic that you calculated and then find the associated p-value, or "t tables" to find the t-statistic associated with a given number of degrees of freedom and confidence level. Although knowing how to read these tables is a useful skill, R has readily available functions that will calculate these values for you, as shown earlier in this chapter.

Correlation versus causation

In Section 6.4, we described an example for which there was a correlation between increased ice cream sales, drowning incidents, and forest fires in the summer but cautioned that none of these events causes the other; heat is most likely the causal mechanism. Although this is a simple example, it illustrates a situation that often happens in data analysis: two or more variables may be correlated, but there might not be causal mechanisms associated with the two variables.

The ability to differentiate correlation and causation has important applications in genomics. Since the completion of the Human Genome Project in 2003, researchers have used various approaches to identify genetic variants that are associated with certain diseases and traits (Genome-wide association studies fact sheet 2020). In one such approach, called the genome-wide association study (GWAS), DNA from people with a particular disease or disorder is compared with DNA from people who do not have the disease. Computers scan the DNA of the two groups for hundreds of thousands of selected genetic markers (called SNPs and pronounced "snips"). If one allele of a SNP or group of SNPs occurs more frequently in the group with the disease, this suggests an association between the SNP alleles (and the gene that is located nearby) and the disease. However, it would not be correct to say that this allele or gene *causes* the disease because GWAS is detecting a correlation only. In fact, many studies have determined that many different genes play a role in initiating the manifestation of a disease or phenotype. Furthermore, a single gene may be associated with more than one disease or trait, and potential interactions between genes and the environment further complicate the causal mechanism. Causal inference is a growing area of statistical research with new models being developed to directly address causal concerns both in experimental design and analyses of observational data (Pearl 2009).

Multiple testing

When conducting a hypothesis test, there is a certain probability that the null hypothesis is rejected even when it is true, which is referred to as a *type I error.* This situation may arise, for example, when we are interested in determining whether many categories are statistically different from each other. The *Bonferroni correction* is often used to adjust the threshold for rejecting the null hypothesis when performing many statistical tests for many hypotheses (Dunn 1961). If the level α is used to test one hypothesis, the adjusted α for rejection of the null hypothesis with the Bonferroni correction for testing the difference between m means (which leads to m tests) is α/m. This correction makes it considerably more difficult to reject the null hypothesis, which protects against type I error.

P-hacking

In 2016, the American Statistical Association (ASA) released a statement on the appropriate use of p-values titled "The ASA Statement on p-values: Context, Process, and Purpose." This was in response to widespread concern about how p-values have been (mis)used inside and outside of the statistical community. Many statisticians and scientists are often concerned with the practice of selectively picking significant or promising findings, referred to as *p-hacking*. For example, some models incorporate randomness, and you might get slightly different results with each model run. An example of p-hacking would be running the model multiple times until you get the p-value you desire. Another example could be subsetting your data to only a certain group of individuals and this subset gives your desired results. Results produced through p-hacking are not reliable and are unethical. To tackle this and other problems with the use of p-values, the ASA has released the following recommended practices and principles:

1. *P-values can indicate how incompatible the data are with a specified statistical model.*

2. *P-values do not measure the probability that the studied hypothesis is true, or the probability that the data were produced by random chance alone.*

3. *Scientific conclusions and business or policy decisions should not be based only on whether a p-value passes a specific threshold.*

4. *Proper inference requires full reporting and transparency.*

5. *A p-value, or statistical significance, does not measure the size of an effect or the importance of a result.*

6. *By itself, a p-value does not provide a good measure of evidence regarding a model or hypothesis.*

(Wasserstein and Lazar 2016)

Check Your Understanding

1. Analyze a dataset that you collected (or one provided by your instructor).

   a. Explore the data, as explained in Section 6.4.

   b. Use a statistical test appropriate for the data, as explained in Section 6.5.

Go to **macmillanlearning.com/knisely6e**
and select **"Student Site"** to access
samples, template files, and tutorial videos

The R code and data for this chapter are posted on Github:
https://github.com/ckelling/data_analysis_using_statistics

## Chapter 7

Revision

Objectives

7.1 Understand that all good writers write multiple drafts before submitting the final version

7.2 Understand that the writing process is a cycle of writing, resting, revising, and rewriting

7.3 Evaluate drafts in passes in which similar elements are evaluated in one pass

7.4 Explain different ways of organizing sentences in paragraphs

7.5 Identify words that make sentences unnecessarily complex (and remove them)

7.6 Differentiate between words that are frequently confused in scientific writing

7.7 Use correct grammar and punctuation to make your writing understandable

7.8 Learn how to give constructive and respectful feedback

7.9 Learn how to use feedback to improve your own writing

7.10 Use the Scientific Writing Checklist to anticipate and avoid common problems in your writing

Revision—reading your paper and making corrections and improvements—is an important task that usually does not get nearly the attention it deserves. Too many students write the first draft of their laboratory report the night before it is due and hand in the hard copy, still warm from the printer, without even having proofread it.

The truth is, most writers cannot produce a clear, concise, and error-free product on the first try. It may take several revisions before the writer is satisfied that he or she has conveyed, with clarity and logic, the motivation for writing the paper, the important findings, and the conclusions. Do not try to write and revise your entire paper in one marathon session. Instead, **break up the writing process** into multiple, shorter segments. The breaks give your mind time to process what you've written. Starting to write early also allows you to get help if necessary, get feedback from your peer reviewer, and make final revisions.

Most excellent writers were not born that way. They achieved excellence through "deliberate practice" (Martin 2011). The old adage "practice

145

makes perfect" applies not just to musicians and athletes, but to *you* as an aspiring author in the biological sciences. So if writing doesn't come naturally to you, take heart. Writing scientific papers becomes easier with practice, especially if you learn from your mistakes.

7.1 Getting Ready to Revise

Take a break

The first step in revision is *not* to do it immediately after you have completed the first draft. You need to distance yourself from the paper to gain the objectivity needed to read the paper critically. So take a break, and go for a run or get a good night's sleep.

Slow down and concentrate

Find a quiet room where you won't be disturbed. Don't read your paper the same way you wrote it. Instead, **change its appearance** by increasing the zoom level or converting the Word document to a PDF (Koerber 2019). Even better, read a printed copy. Next, use some of your other senses to force yourself to **slow down**. For example, reading aloud involves the sense of hearing. Pointing to each word with your finger adds the sense of touch. If something doesn't sound right, trust your instincts. Figure out what is bothering you and fix the troublesome passage. Finally, **don't try to fix every kind of error in one pass**. If you do, you're sure to miss some.

Think of your audience

The rest of this chapter describes a systematic approach to revising your writing, whether for a lab report, a science communication for a general audience, a poster, or an oral presentation. Remember for whom you are writing and keep in mind the needs and motivations of your audience. Revise your writing with the goal of meeting or exceeding their expectations in a comprehensible style.

7.2 Editing

Revision can be divided into two stages: editing and proofreading. **Editing** is done first and involves reading for content and organization. The editing process proceeds from the broad to the specific. First, evaluate the overall structure of your paper. Then, read each individual section, paragraph, sentence, and word critically. And don't forget to check that the data were plotted correctly and that the description of each visual in the text is accurate. After editing is completed, **proofread** the paper. This involves correcting errors in spelling, punctuation, grammar, and overall format. This chapter provides specific guidance for each step of the revision process. Section 8.2 illustrates common errors to look out for when you revise your written work.

Evaluate the overall structure

If your instructor provided a rubric for your assignment, use it as a checklist for content and organization. You may also wish to print out the "Scientific Writing Checklist" from **macmillanlearning.com/knisely6e**. If you are preparing a paper for submission to a journal or conference, follow the relevant *Instructions to Authors*.

Most scientific papers are divided into **standard sections**: title, authors' name(s), abstract, *Introduction, Materials & Methods, Results, Discussion,* and references (known as IMRD format). Readers of scientific papers like this format because they know where to look to find certain kinds of information. Confirm that these sections are in the **right order** in your paper. Then check that each section has the **appropriate content** (Table 7.1) by underlining each component on the printed pages or highlighting them on the computer screen (Hofmann 2019).

Do the math at least twice

Double-check your calculations and spreadsheet data entries. A mistake at this stage will have a negative domino effect, resulting in inaccurate figures or tables and a faulty discussion and interpretation of the results.

Organize each section

The phrases you underlined when you checked for content should provide a rough outline of each section. Do the **most important topics stand out**? Does the **order of the topics make sense** chronologically or sequentially? Is the order what your audience expects (for example, are the topics arranged from broad to specific in the Introduction section and specific to broad in the Discussion section)? Rearrange paragraphs so that the important topics can be identified easily, in an order that makes sense.

Make coherent paragraphs

Each paragraph should focus on only one topic   Make the topic sentence the first sentence in the paragraph. Follow the topic sentence with supporting sentences that directly relate to the main idea.

TABLE 7.1	Checklist for section content
Section	**Content**
Title	Contains keywords that describe the essence of the study.
	"Filler" phrases like "Analysis of the...," or "Study of the..." are not used.
Abstract	Contains an introduction, a brief description of the methods, results, and conclusions.
Introduction	The structure is broad to specific: The main topic is introduced on a general level. The question or unresolved problem is stated. The objectives of the current study are presented in the last paragraph.
	Background information is provided by citing published sources.
Materials and Methods	Contains all of the relevant details to enable another trained scientist to repeat the procedure.
	Routine procedures are not described.
Results	Contains text and visuals. The descriptive text precedes each visual and includes a reference to the figure or table number.
	The text describes the results shown in each visual. The results are not explained or interpreted.
	Visuals include graphs, tables, photos, gel images, and so on, which contain the numerical or descriptive data. Each visual has its own caption that begins with Figure (or Table) and a number followed by a title that can be understood apart from the text.
Discussion	The structure is specific to broad: The results are explained and interpreted. The results are compared to those in other studies, usually published journal articles. If warranted, there may be a discussion of why the results in the current study are important or how these results contribute to our understanding of the topic.
References	The full reference is given for each source cited in the text.
	References that have not been cited are not listed.

Arrange the sentences in a logical order   Different strategies can be used depending on the section of the scientific paper. For example, in the Materials and Methods section, it makes sense to describe the procedure *sequentially* (in the order the steps were carried out).

> EXAMPLE: Barley seeds were surface sterilized with 10% bleach. Then they were cut in half, keeping the endosperm portion and discarding the embryo portion. Each half was placed cut side down on three pieces of sterile filter paper that had been soaked in 3.5 mL of HEPES-EGTA-Ca^{2+} buffer or a certain concentration of hormone solution.

Chronological sentence order might be used in the Introduction section to describe the sequence of events leading up to our current state of knowledge about a topic.

EXAMPLE: Germination begins when a seed imbibes water. Early studies on barley seeds showed that the hormone gibberellic acid (GA) is involved in this process (Paleg 1960; Yomo 1960). Varner (1964) demonstrated that the enzymes that digest starch into sugars are produced in the aleurone layer. For the past 5 decades, it was suspected that GA binds to a receptor on the surface of the aleurone cells, but how the protein subunits in the receptor function in signal transduction was discovered only recently (Ueguchi-Tanaka et al. 2005).

General-to-specific is another way to arrange sentences in the introduction. In this approach, the paragraph starts with a general idea that is then supported by details or examples.

EXAMPLE: Germination begins when a seed imbibes water. Water triggers the release of gibberellic acid (GA) from the embryo. The hormone diffuses through the endosperm and binds to receptors on the plasma membrane of the aleurone cells. Through a signal transduction pathway that is not fully understood, digestive enzymes are activated and released into the endosperm. One of those enzymes is α-amylase, which metabolizes starch into its sugar subunits.

On the other hand, *specific-to-general* is the order expected in the Discussion section. As shown in the following example, the paragraph begins with a brief recapitulation of a specific result. Subsequent sentences explain the result based on our current understanding of the topic.

EXAMPLE: Embryoless half-seeds exposed to higher concentrations of gibberellic acid had a lower percentage of starch remaining after one week (Figure 2). Gibberellic acid is known to induce the production of α-amylase, an enzyme that hydrolyzes starch (Lovegrove and Hooley 2000). During germination, gibberellic acid, produced by the embryo, binds to receptors in the aleurone layer and activates α-amylase, which then degrades the starch in the endosperm into glucose, which provides the energy for the embryo to grow.

Use signal words and phrases to guide readers from one sentence to the next
Signal words and phrases (also called **transitions**) help readers see relationships between sentences. Smooth transitions help readers to follow the writer's thought process, thereby increasing comprehension.

Consider the following example:

> FAULTY: Catalase is an enzyme that breaks down hydrogen peroxide in both plant and animal cells. Low or high temperature can lower the rate at which the catalase can react with the hydrogen peroxide. In optimal conditions, the enzyme functions at a rate that will prevent any substantial buildup of the toxin. If the temperature is too low, the rate will be too slow, but high temperatures lead to the denaturation of the enzyme.

Where is the writer going with this paragraph? The sentences do not seem to flow because there is no guidance from the writer on how one sentence is related to the next. To improve flow, use signal words and phrases such as *however, thus, although, in contrast, similarly, on the other hand, in addition to*, and *furthermore*. Signal words and phrases may also be key words that are repeated from one sentence to the next. Notice how the addition of transitions guides the reader step by step through this passage.

> REVISION: Catalase is an enzyme that breaks down hydrogen peroxide in both plant and animal cells. One of the factors that affects the rate *of this reaction* is temperature. At optimal *temperatures*, the rate is sufficient to prevent substantial buildup of the toxic *hydrogen peroxide*. If the temperature is too low, *however*, the rate will be too slow, and hydrogen peroxide *accumulates* in the cell. *On the other hand*, high temperatures may denature the *enzyme*.

Write meaningful sentences

Each sentence should say something meaningful and not repeat what was said before (avoid redundancy). Consider the following examples.

EXAMPLE: ~~From the data that has been gathered, a graph depicting the effect of various pH environments on the rate of catalase activity is represented. The graph displays the pH tested and the reaction rate. The data plotted is an accumulation of data from several lab sections. Through analyzing the graph I can see that T~~there ~~is~~ was no activity below a pH of 4 or above 10. Maximum catalase activity occurred at pH 7.

EXPLANATION: The first three sentences do not convey anything substantive about the *results*. Only the last two sentences contain meaningful information.

EXAMPLE: As the enzyme concentration increased, the initial velocity increased ~~as well~~ (Figure 3). ~~There is an overall gradual increasing relationship between the enzyme concentration and the initial velocity. This increasing relationship seems to remain constant from the lowest concentration of 0 mM to the highest concentration which is 2 mM. This graph showed that there was a fairly increasing positive linear relationship between enzyme concentration and initial velocity.~~

EXPLANATION: The last three sentences say the same thing as the first sentence. The only substantive piece of information missing from the first sentence is that the trend was linear. Eliminate repetition and describe the results in as few words as possible.

To edit redundant sentences, take the best parts of those sentences and combine them into one concise sentence. Put yourself in your reader's shoes: Would you rather waste precious minutes wading through verbiage or get needed information with minimal effort?

Technical accuracy Sentences that provide background information on a topic (as in the Introduction section), describe procedures (in the Materials and Methods section), or explain results (in the Discussion section) should be based on scientific fact. When in doubt, check your references, including secondary sources such as your textbook. Furthermore, make sure that your description of the results shown in each visual is accurate. In particular, pay attention to words like *increase* and *decrease*. Check that you did not mix up the results when you describe multiple data sets plotted on one graph.

Sentence length   Short sentences that contain only one idea are easy to understand. A text that contains nothing but short sentences, however, may be perceived as childish at best or hard to follow at worst. On the other hand, long, needlessly complex sentences obscure the main idea and slow comprehension. Aim for a mixture of short and long sentences in your writing. Use more words to explain a complex idea, but keep each sentence focused on just one idea.

Here are some examples of **needlessly complex sentences**.

FAULTY 1:   *There are* two protein assays *that* are often used in research laboratories.

REVISION: Two protein assays are often used in research laboratories. (Avoid unnecessary words and phrases that "pad" a sentence.)

FAULTY 2:   *It is interesting to note that* some enzymes are stable at temperatures above 60 °C.

REVISION: Some enzymes are stable at temperatures above 60 °C. (Avoid unnecessary introductory phrases.)

FAULTY 3:   *The analyses were done on* the recombinant DNA to determine which piece of foreign DNA was inserted into the vector.

REVISION: The recombinant DNA was analyzed to determine which piece of foreign DNA was inserted into the vector. (Make *DNA*, not the *analyses*, the subject of the sentence.)

FAULTY 4:   *We make the recommendation* that micropipettors be used to measure volumes less than 1 mL.

REVISION: We recommend that micropipettors be used to measure volumes less than 1 mL. (Replace sluggish noun phrases [nominalizations] with verb phrases.)

FAULTY 5:   These assays alone cannot *tell* what the protein concentration of a substance is.

REVISION: These assays alone cannot determine the protein concentration of a substance. (Replace colloquial expressions with precise alternatives.)

Emphasize the subject   Putting the subject at the beginning of the sentence makes the subject stand out. Position the verb close by so that there is no doubt about the subject's action.

Active and passive voice   In **active voice**, the subject *performs* the action. In **passive voice**, the subject *receives* the action. Consider the following example:

PASSIVE: The clam was opened by the sea star. (Emphasis on *clam*)

ACTIVE: The sea star opened the clam. (Emphasis on *sea star*)

Although the meaning is the same in both sentences, notice the difference in emphasis. In active voice, the emphasis is on the performer, and the action takes place in the direction the reader reads the sentence. Active voice is recommended by most style guides for reasons that include the following:

- It sounds more natural and is easier for the reader to process.
- It is shorter and more dynamic.
- There is no ambiguity about who/what the subject of the sentence is, or about who did the action.

Consider the following example:

PASSIVE: It was concluded from this observation that…

ACTIVE: I concluded from this observation that…

Passive voice leaves the reader wondering who drew the conclusion; active voice conveys this information clearly.

While active voice is generally preferred, passive voice may be more appropriate when *what* is being done is more important than *who* is doing it. For example:

PASSIVE: Catalase was extracted from a turnip. (Emphasis on *catalase*)

ACTIVE: I extracted catalase from a turnip.
(Emphasis on *I*)

Notice the difference in emphasis. Is it really important to the success of the procedure that *you* did it, or does the emphasis belong on the catalase?

A paper that contains a mixture of active and passive voice is pleasant to read. Your decision to use active or passive voice in a sentence should ultimately be determined by clarity and brevity. In other words, use active voice to emphasize the subject and the fact that the subject is performing the action. Use passive voice when the action is more important than who is doing it.

Present or past tense   In scientific papers, present tense is used mainly in the following situations:

- To make generally accepted statements (for example, "Photosynthesis *is* the process whereby green plants produce sugars").
- When referring directly to a table or figure in your paper (for example, "Figure 1 *is* a schematic diagram of the apparatus").
- When stating the findings of published authors (for example, "Catalase HPII from *E. coli is* highly resistant to denaturation [Switala and others 1999]").

Past tense is used mainly in the following situations:

- To report your own work, especially in the abstract, Materials and Methods, and Results sections, because it remains to be seen if the scientific community accepts your work as fact (for example, "At temperatures above 37 °C, catalase activity *decreased* (Figure 3)").
- To cite another author's findings directly (for example, "Miller and others (1998) *found* that...").

Choose your words carefully

Words are the basic organizational unit of language. The words you choose and how you arrange them in a sentence will determine how well you convey your message to your readers. Beware of the following word-level problems:

Keep related words together   Consider the following sentence taken from an English-language newspaper in Japan: "A committee was formed to examine brain death in the Prime Minister's office." Although brain death in the Prime Minister's office may be a political reality, what was really intended was, "A committee was formed in the Prime Minister's office to examine brain death."

TABLE 7.2 Examples of redundancy	
Redundant	Revised
It is absolutely essential…	It is essential…
mutual cooperation	cooperation
basic fundamental concept	basic concept or fundamental concept
totally unique	unique
The solution was obtained and transferred…	The solution was transferred…

Redundancy   **Redundancy** means using two or more words that mean the same thing. This problem is easily corrected by eliminating one of the redundant words (Table 7.2). Along with empty phrases, redundancy is a source of **wordiness**, using too many words to convey an idea.

Some people think that using more words makes them sound important. In science, however, wordiness should be avoided at all costs, because it indicates that the writer can't communicate clearly. For student writers whose papers are evaluated by instructors, lack of clarity translates into a low grade. Researchers and faculty members, whose reputation depends on the number and quality of their publications, simply cannot afford *not* to write clearly, because poorly written papers may be equated with shoddy scientific methods.

Empty phrases   Replace empty phrases with a concise alternative (Table 7.3). Put yourself in your reader's shoes. Which of the following two sentences (inspired by VanAlstyne 2005) would you rather read?

FAULTY: It is absolutely essential that you use a minimum number of words in view of the fact that your reader has numerous other tasks to complete at the present time.

REVISION: Write concisely, because your reader is busy.

Initially it is difficult to write in (and read) the terse, get-to-the-point style that characterizes scientific papers. With practice, however, you may come to appreciate this style because in a well-written paper, not a word is wasted. The benefit to you as a reader is that you extract a maximum amount of information from a minimum amount of text.

Ambiguous use of *this*, *that*, and *which*   Ambiguity results when *this*, *that*, or *which* could refer to more than one subject.

TABLE 7.3 Examples of empty phrases

Empty	Concise
a downward trend	a decrease
a great deal of	much higher
a majority of	most
accounted for the fact that	because
as a result	so, therefore
as a result of	because
as soon as	when
at which time	when
at all times	always
at a much greater rate than	faster
at the present time, at this time	now, currently
based on the fact that	because
brief in duration	short, quick
by means of	by
came to the conclusion	concluded
despite the fact that, in spite of the fact that	although, though
due to the fact that, in view of the fact that	because
for this reason	so
in fact	*omit this phrase*
functions to, serves to	*omit this phrase*
degree of	higher, more
in a manner similar to	like
in the amount of	of
in the vicinity of	near, around
is dependent upon	depends on
is situated in	is in
it is interesting to note that, it is worth pointing out that	*omit these kinds of unnecessary introductions*
it is recommended	I (we) recommend
on account of	because, due to
prior to	before
provided that	if
referred to as	called
so as to	to
through the use of	by, with
with regard to	on, about
with the exception of	except
with the result that	so that

FAULTY: The data show that the longer the enzyme was exposed to the salt solution, the lower the enzyme activity in the assay. *This* means that the salt changes the conformation of the enzyme, *which* makes it less reactive with the substrate.

EXPLANATION: The subject of *this* and *which* is unclear.

REVISION: The longer the enzyme was exposed to the salt solution, the lower the enzyme activity in the assay. Exposure to the salt solution may change the conformation of the enzyme, resulting in lower enzyme-substrate activity.

Ambiguous use of pronouns (him, her, it, he, she, its)   Ambiguity results when a pronoun could refer to two possible antecedents.

FAULTY: With time, salt changes the conformation of the enzyme, which makes *it* less reactive with the substrate.

EXPLANATION: *It* could refer to *salt* or *enzyme*. To eliminate the ambiguity, replace *it* with the appropriate noun phrase.

REVISION: With time, salt changes the conformation of the enzyme, so that the enzyme can no longer react with its substrate.

Word usage   When you use the right words in the right situations, readers have confidence in your work. Use a standard dictionary whenever you are not sure about word usage. Consult your textbook and laboratory exercise for proper spelling and usage of technical terms. The following word pairs are frequently confused in students' scientific papers.

absorbance, absorbency, observance   *Absorbance* is how much light a solution absorbs; absorbance is measured with a spectrophotometer. *According to Beer's law, absorbance is proportional to concentration. Absorbency* is how much moisture a diaper or paper towel can hold. *Brand A paper towels show greater absorbency than Brand B paper towels. Observance* is the act of observing. *Government offices are closed today in observance of Independence Day.*

affect, effect   *Affect* is a verb that means "to influence." *Temperature affects enzyme activity. Affect* is rarely used as a noun in biology, although it has a specific meaning in psychology. *Effect* can be used either as a noun or a verb. When used as a noun, *effect* means "result." *We studied the effect of temperature on enzyme activity.* When used as a verb, *effect* means "to cause." *High temperature effected a change in enzyme conformation, which destroyed enzyme activity.*

alga, algae   See plurals.

amount, number   Use *amount* when the quantity cannot be counted. *The reaction rate depends on the amount of enzyme in the solution.* Use *number* if you can count individual pieces. *The reaction rate depends on the number of enzyme molecules in the solution.*

analysis, analyses   See plurals.

bacterium, bacteria   See plurals.

bind, bond   *Bind* is a verb meaning "to link." *The active site is the region of an enzyme where a substrate binds. Bond* is a noun that refers to the chemical linkage between atoms. *Proteins consist of amino acids joined by peptide bonds. Bond* used as a verb means "to stick together." *This 5-minute epoxy glue can be used to bond hard plastic.*

complementary, complimentary   *Complementary* means "something needed to complete" or "matching." *The DNA double helix consists of complementary base pairs: A always pairs with T, and G with C. Complimentary* means "given free as a courtesy." *The brochures at the visitor's center are complimentary.*

confirmation, conformation   *Confirmation* means "verification." *I received confirmation from the postal service that my package had arrived. Conformation* is the three-dimensional structure of a macromolecule. *Noncovalent bonds help maintain a protein's stable conformation.*

continual, continuous   *Continual* means "going on repeatedly and frequently over a period of time." *The continual chatter of a group of inconsiderate students during the lecture annoyed me. Continuous* means "going on without interruption over a period of time." *The bacteria were grown in L-broth continuously for 48 hr.*

create, prepare, produce   *Create* is to cause to come into existence. *The artist used wood and plastic to create this sculpture. Prepare* means "to make ready." *The protein standards were prepared from a 50 mg/mL stock solution. Produce* means to make or manufacture. *The reaction between hydrogen peroxide and catalase produces water and oxygen.*

datum, data   See plurals.

different, differing   *Different* is an adjective that means "not alike." An adjective modifies a noun. *Different concentrations of bovine serum albumin were prepared. Differing* is the intransitive tense of "to differ," a verb that means "to vary." It is incorrect to replace the word *different* with *differing* in the preceding example, because *differing* implies that a single concentration changes depending on time or circumstance. This situation is highly unlikely with bovine serum albumin, which is quite stable under laboratory conditions! An acceptable use of *differing* is shown in the following example: *Bovine serum albumin solutions, differing in their protein content, were prepared.*

effect, affect   See affect, effect.

fewer, less   Use *fewer* when the quantity can be counted. *The reaction rate was lower, because there were fewer collisions between enzyme and substrate molecules.* Use *less* when the quantity cannot be counted. *The weight of this sample was less than I expected.*

formula, formulas, formulae   See plurals.

hypothesis, hypotheses   See plurals.

its, it's   *Its* is a possessive pronoun meaning "belonging to it." *The Bradford assay is preferred because of its greater sensitivity. It's* is a contraction of "it is." *The Bradford assay is preferred because it's more sensitive.* (Note: Contractions should not be used in formal writing.)

less, fewer   See fewer, less.

lose, loose   *Lose* means to misplace or fail to maintain something. *An enzyme may lose its effectiveness at high temperatures. Loose* means "not tight." *When you autoclave solutions, make sure the lid on the bottle is loose.*

lowered, raised Both of these are transitive verbs, which means that they require a direct object (a noun to act on). **Wrong**: *The fish's body temperature lowered in response to the cold water.* **Right**: *The cold water lowered the fish's body temperature.*

media, medium See plurals.

observance See absorbance, absorbency, observance.

phenomenon, phenomena See plurals.

plurals The plural and singular forms of some words used in biology are given in Table 7.4. A common mistake with these words is not making the subject and verb agree. Some disciplines treat *data* as singular, but scientists subscribe to the strict interpretation that *data* is plural. The data *show*... (not *shows*) is correct.

prepare See create, prepare, produce.

produce See create, prepare, produce.

raised, lowered See lowered, raised.

ratio, ration *Ratio* is a proportion or quotient. *The ratio of protein in the final dilution was 1:5.* *Ration* is a fixed portion, often referring to food. *The Red Cross distributed rations to the refugees.*

TABLE 7.4 Singular and plural of words frequently encountered in biology

Singular	Plural
alga	algae
analysis	analyses
bacterium	bacteria
criterion	criteria
datum (rarely used)	data
formula	formulas, formulae
hypothesis	hypotheses
index	indexes, indices
medium	media
phenomenon	phenomena
ratio	ratios

strain, strand   A *strain* is a line of individuals of a certain species, usually distinguished by some special characteristic. *The lacI⁻ strain of E. coli produces a nonfunctional repressor protein.* A *strand* is a ropelike length of something. *The strands of DNA are held together with hydrogen bonds.*

than, then   *Than* is an expression used to compare two things. *Collisions between molecules occur more frequently at high temperatures than at low temperatures. Then* means "next in time." *First 1 mL of protein sample was added to the test tube. Then 4 mL of biuret reagent was added.*

that, which   Use *that* with restrictive clauses. A restrictive clause limits the reference to a certain group. Use *which* with nonrestrictive clauses. A nonrestrictive clause does not limit the reference, but rather provides additional information. Commas are used to set off nonrestrictive clauses but not restrictive clauses. Consider the following examples:

EXAMPLE 1:   The Bradford assay, which is one method for measuring protein concentration, requires only a small amount of sample. (*Which* begins a phrase that provides additional information, but is not essential to make a complete sentence.)

EXAMPLE 2:   Enzyme activity decreased significantly, which suggests that the enzyme was denatured at 50 °C. (*Which* refers to the entire phrase *Enzyme activity decreased significantly,* not to any specific element.)

EXAMPLE 3:   The samples that had high absorbance readings were diluted. (*That* refers specifically to *the samples.*)

various, varying   *Various* is an adjective that means "different." *Various hypotheses were proposed to explain the observations. Varying* is a verb that means "changing." *Varying the substrate concentration while keeping the enzyme concentration constant allows you to determine the effect of substrate concentration on enzyme activity.* Analogous to *different, differing,* replacing the word *various* with *varying* in the preceding example changes the meaning of the sentence. *Varying* implies that a single hypothesis changes depending on time or circumstance. *Various* implies that different hypotheses were proposed.

Jargon and scientific terminology   Jargon refers to words and abbreviations used by specialists. Whenever you use terms that may be unfamiliar to your audience, define them. Always write out the full expression when first using an abbreviation. Words that you learned in class are *not* jargon and should be used in your scientific writing. When you use scientific terminology correctly, your readers have confidence in your knowledge. On the other hand, when you are communicating with a general audience, do not use jargon. Instead, use language that non-specialist readers can understand.

Colloquial language   In general, colloquial language should not be used in formal, scientific writing. However, science communications for a general audience tend to be less formal, and colloquial expressions may be used effectively to explain concepts and generate interest in the topic.

Gender-neutral language   Years ago, it was customary, for the sake of simplicity, to use masculine pronouns to refer to antecedents that could be masculine or feminine, but that use of language is no longer accepted.

> SEXIST: The clarity with which a biology student
> writes *his* lab reports affects *his* grade.

This practice is no longer considered to be politically correct. One solution that preserves equality, but makes sentences unnecessarily complex, is to include both masculine and feminine pronouns, as in the following example.

> EQUAL BUT   The clarity with which a biology student
> AWKWARD: writes *his or her* lab reports affects *his or
> her* grade.

Two better alternatives are to make the antecedent plural (revision 1) or to rewrite the sentence to avoid the gender issue altogether (revision 2).

> REVISION 1:   The clarity with which biology students write
> *their* lab reports affects *their* grade.

> REVISION 2:   *Writing clearly* has a positive effect on a biology
> student's grade. (Change the subject from
> *biology* student to *writing clearly*.)

Construct memorable visuals

Visuals often make the difference in how well you convey your message to your readers or listeners. Make sure you use the **appropriate visual** for the data (see Section 5.4). Make sure **every visual serves a purpose**, because unnecessary visuals only dilute the significance of your message. Check that the **visuals are positioned in the right order** and that **each visual is described** in the text.

7.3 Proofreading: The Home Stretch

Proofreading is the last stage of revision. Like editing, it requires intense focus and slow, careful reading to find errors in format, spelling, punctuation, and grammar. **Grammar** refers to the rules that deal with the form and structure of words and their arrangement in sentences. See Hacker and Sommers (2021), Bullock et al. (2021), or Lunsford (2019, 2020, 2021) for a more comprehensive treatment of the subject.

Make subjects and verbs agree

We learn early on in our formal education to make the verb agree with the subject. Most of us know that *the sample was...,* but that *the samples were....* Most errors with subject–verb agreement occur when there are words *between* the subject and the verb, as in the following example.

> EXAMPLE: The *kinetic energy* of molecules *is* (not *are*) lower at 6 °C than 45 °C.

> EXAMPLE: The *chance* of collisions between enzyme and substrate molecules *increases* (not *increase*) under those conditions.

> EXAMPLE: The enzyme has a *range* of temperatures that *is* (not *are*) optimal for activity.

When you write complex sentences, ask yourself what the subject of the sentence is. Look for the verb that goes with that subject. Then, mentally remove the words in between the two, and make the subject and its verb agree.

A second situation in which subject–verb agreement becomes confusing is when there are two subjects joined by *and,* as in the following example.

> EXAMPLE: An enzyme's amino acid *sequence and* its three-dimensional *structure make* (not *makes*) the enzyme–substrate relationship unique.

Compound subjects joined by *and* are almost always plural.

A third situation involving subject–verb agreement is that when numbers are used in conjunction with units, the *quantity* is considered to be *singular*, not plural.

> EXAMPLE: To extract the enzyme, 12 g of turnips *was* (not *were*) homogenized with 150 mL of cold, distilled water.

Write in complete sentences

A complete sentence consists of a subject and a verb. If the sentence starts with a subordinate word or words such as *after, although, because, before, but, if, so that, that, though, unless, until, when, where, who,* or *which,* however, another clause must complete the sentence.

FAULTY 1: High temperatures destroy the three-dimensional structure of enzymes. Thus changing the effectiveness of the enzymes. (The second "sentence" is a fragment.)

REVISION: High temperatures destroy the three-dimensional structure of enzymes, thus changing their effectiveness. (Combine the fragment with the previous sentence, changing punctuation as needed.)

FAULTY 2: Sessile organisms typically exhibit radial symmetry. Although most motile organisms are bilaterally symmetric. (The second "sentence" is a fragment.)

REVISION: Sessile organisms typically exhibit radial symmetry. Most motile organisms are bilaterally symmetric. (Delete the subordinate word[s] to make a complete sentence.)

Revise run-on sentences

Run-on sentences consist of two or more independent clauses joined without proper punctuation. Each independent clause could stand alone as a complete sentence. Run-on sentences are common in first drafts, where your main objective is to get your ideas down on paper (or electronic media, if you use a computer). When you revise your first draft, however, use one of the following strategies to revise run-on sentences:

- Insert a comma and a coordinating conjunction (*and, but, or, nor, for, so,* or *yet*).
- Use a semicolon or possibly a colon.
- Make two separate sentences.
- Rewrite the sentence.

FAULTY 1: Most animals have 2 or 3 tissue layers, sponges have none.

REVISION A: Most animals have 2 or 3 tissue layers, but sponges have none. (Use a coordinating conjunction.)

REVISION B: Most animals have 2 or 3 tissue layers; sponges have none. (Use a semicolon.)

FAULTY 2: Plants alternate between a multicellular diploid stage and a multicellular haploid stage, this life cycle is called alternation of generations. (Fused sentence.)

REVISION A: Plants alternate between a multicellular diploid stage and a multicellular haploid stage; this life cycle is called alternation of generations. (Use a semicolon to separate the two clauses.)

REVISION B: Plants alternate between a multicellular diploid stage and a multicellular haploid stage. This life cycle is called alternation of generations. (Make two separate sentences.)

FAULTY 3: An increase in enzyme concentration increased the reaction rate as did an increase in substrate concentration, so the concentrations of the molecules have an influence on how the enzyme reacts.

REVISION A: As enzyme concentration and substrate concentration increased, so did the reaction rate. (Rewrite the sentence. The second half of the original sentence was deleted because it says nothing meaningful.)

REVISION B: Enzyme and substrate concentration influence enzyme reaction rate: an increase in enzyme or substrate concentration increased reaction rate. (Use a colon.)

Spelling

Spell checkers in word processing programs are so easy to use that there is really no excuse for *not* using them. Just remember that spell checkers may not know scientific terminology, so consult your textbook or laboratory manual for correct spelling. In some cases, the spell checker may even try to get you to change a properly used scientific word to an inappropriate word that happens to be in its database (for example, *absorbance* to *absorbency*).

The following poem is an example of how indiscriminate use of the spell checker can produce garbage:

Wrest a Spell

Eye halve a spelling chequer
It came with my pea sea
It plainly marques four my revue
Miss steaks eye kin knot sea.

Eye strike a key and type a word
And weight four it two say
Weather eye am wrong oar write
It shows me strait a weigh.

As soon as a mist ache is maid
It nose bee fore two long
And eye can put the error rite
Its rare lea ever wrong.

Eye have run this poem threw it
I am shore your pleased two no
Its letter perfect awl the weigh
My chequer tolled me sew.

— Sauce unknown

Spell checkers will also not catch mistakes of usage, for example *form* if you really meant *from*. Print out your document and proofread the hard copy carefully.

Punctuation

The purpose of punctuation marks is to divide sentences and parts of sentences to make the meaning clear. A few of the most common punctuation marks and their uses are described in this section. For a more comprehensive, but still concise, treatment of punctuation, see Hacker and Sommers (2021) or Lunsford (2019, 2020, 2021).

The comma   The comma inserts a pause in the sentence in order to avoid confusion. Note the ambiguity in the following sentence:

While the sample was incubating the students prepared the solutions for the experiment.

A comma *should* be used in the following situations:

1. To connect two independent clauses that are joined by *and, but, or, nor, for, so,* or *yet.* An independent clause contains a subject and a verb, and can stand alone as a sentence.

   > EXAMPLE: Feel free to call me at home, but don't call after 9 p.m.

2. After an introductory clause, to separate the clause from the main body of the sentence.

   > EXAMPLE: Although she spent many hours writing her lab report, she earned a low grade because she didn't answer all of the questions.

   A comma is not needed if the clause is short.

   > EXAMPLE: Suddenly, the power went out.

3. Between items in a series, including the last two.

   > EXAMPLE: Enzyme activity is affected by factors such as substrate concentration, pH, temperature, and salt.

4. Between coordinate adjectives (if the adjectives can be connected with *and*).

   > EXAMPLE: The students' original, humorous remarks made my class today particularly enjoyable. (*Original and humorous remarks* makes sense.)

5. A comma is not needed if the adjectives are cumulative (if the adjectives cannot be connected with *and*).

   > EXAMPLE: The three, tall students look like football players. (It would sound strange to say *three and tall students.*)

6. With *which,* but not *that* (see Word usage; that, which,)

7. To set off conjunctive adverbs such as *however, therefore, moreover, consequently, instead, likewise, nevertheless, similarly, subsequently, accordingly,* and *finally.*

   > EXAMPLE: Instructors expect students to hand in their work on time; however, illness and personal emergencies are acceptable excuses.

8. To set off transitional expressions such as *for example, as a result, in conclusion, in other words, on the contrary,* and *on the other hand.*

> EXAMPLE: Chuck participates in many extracurricular activities in college. As a result, he rarely gets enough sleep.

9. To set off parenthetical expressions. Parenthetical expressions are statements that provide additional information; however, they interrupt the flow of the sentence.

> EXAMPLE: Fluency in a foreign language, as we all know, requires years of instruction and practice.

A comma *should not* be used in the following situations.

1. After *that,* when *that* is used in an introductory clause

> EXAMPLE: The student could not believe that he lost points on his laboratory report because of a few spelling mistakes.

2. Between cumulative adjectives, which are adjectives that would not make sense if separated by the word *and* (see Item 4 in preceding list)

The semicolon   The semicolon inserts a stop between two independent clauses not joined by a coordinating conjunction (*and, but, or, nor, for, so,* or *yet*). Each independent clause (one that contains a subject and a verb) could stand alone as a sentence, but the semicolon indicates a closer relationship between the clauses than if they were written as separate sentences.

> EXAMPLE: Outstanding student-athletes use their time wisely; this trait makes them highly sought after by employers.

A semicolon is also used to separate items in a series in which the items are already separated by commas.

> EXAMPLE: Participating in sports has many advantages. First, you are doing something good for your health; second, you enjoy the camaraderie of people with a common interest; third, you learn discipline, which helps you make effective use of your time.

The colon   The colon is used to call attention to the words that follow it. Some conventional uses of a colon are shown in the following examples.

Dear Sir or Madam:

5:30 p.m.

2:1 (ratio)

In references, to separate the place of publication and the publisher, as in

Sunderland (MA): Sinauer Associates, Inc.

A colon is often used to set off a list, as in the following example.

> **EXAMPLE:** Catalase activity has been found in the following vegetables: turnips, leeks, parsnips, onions, zucchini, carrots, and broccoli.

A colon *should not* be used when the list follows the words *are, consist of, such as, including,* or *for example.*

> **EXAMPLE:** Catalase activity has been found in vegetables such as turnips, leeks, parsnips, onions, zucchini, carrots, and broccoli.

The period   The period is used to end all sentences except questions and exclamations. It is also used in American English for some abbreviations, for example, *Mr., Ms., Dr., Ph.D., i.e.,* and *e.g.*

Parentheses   Parentheses are used mainly in two situations in scientific writing: to enclose supplemental material and to enclose references to visuals or sources. Use parentheses sparingly because they interrupt the flow of the sentence.

> **EXAMPLE:** Human error (failure to make the solutions correctly, arithmetic errors, and failure to zero the spectrophotometer) was the main reason for the unexpected results.

> **REFERENCE TO VISUAL:** There was no catalase activity above 70 °C (Figure 1).

> **CITATION-SEQUENCE SYSTEM:** C-fern spores do not germinate in the dark (1).

> **NAME-YEAR SYSTEM:** C-fern spores do not germinate in the dark (Cooke and others 1987).

The dash   The dash is used to set off material that requires special emphasis. To make a dash on the computer, type two hyphens without a space before, after, or in between. In some word processing programs, the two hyphens are automatically converted to a solid dash.

Similar to commas and parentheses, a pair of dashes may be used to set off supplemental material.

> EXAMPLE: Human error--failure to make the solutions correctly, arithmetic errors, and failure to zero the spectrophotometer--was the main reason for the unexpected results. (If the word processing program has been set up to convert the two hyphens to a solid dash, the sentence looks like this: Human error—failure to make… spectrophotometer—was the main reason…)

Similar to a colon, a single dash calls attention to the information that follows it.

> EXAMPLE: Catalase activity has been found in many vegetables—turnips, leeks, parsnips, onions, zucchini, carrots, and broccoli.

If an abrupt or dramatic interruption is desired, use a dash. If the writing is more formal or the interruption should be less conspicuous, use one of the other three punctuation marks. However, do not replace a pair of dashes with commas when the material to be set off already contains commas, as in the following example.

> EXAMPLE: The instruments that she plays—oboe, guitar, and piano—are not traditionally used in the marching band.

Abbreviations

The CSE Manual (2014) defines standard abbreviations for authors and publishers in the sciences and mathematics. Some of the terms and abbreviations that you may encounter in introductory biology courses are shown in Table 7.5. Take note of spacing, case (capital or lowercase letters), and punctuation use. Except where noted, the symbols are the same for singular and plural terms (for example, 30 min *not* 30 mins).

Widely known abbreviations such as DNA and ATP do not have to be defined. But abbreviations known only to specialists should be defined the first time they are used.

> EXAMPLE: CRISPR (clustered regularly interspaced short palindromic repeats) technology makes it possible to edit segments of DNA in a precise and predictable fashion.

TABLE 7.5 Standard abbreviations in scientific writing

Term	Symbol or Abbreviation	Example
Latin words and phrases [The CSE Manual (2014) recommends that Latin words be replaced with English equivalents.]		[The Latin word may be replaced with the English equivalent given in brackets.]
circa (approximately)	ca.	The lake is ca. [approx.] 300 m deep.
et alii (and others)	et al.	Jones et al. [and others] (1999) found that …
et cetera (and so forth)	etc.	pH, alkalinity, etc. [and other characteristics] were measured.
exempli gratia (for example)	e.g.	Water quality characteristics (e.g., [for example,] pH, alkalinity) were measured.
id est (that is)	i.e.	The enzyme was denatured at high temperatures, i.e., the enzyme activity was zero. [Because the enzyme was denatured at high temperatures, the enzyme activity was zero.]
nota bene (take notice)	NB	NB [Important!]: Never add water to acid when making a solution.
LENGTH		
nanometer (10^{-9} meter)	nm	*Note:* There is a space between the number and the abbreviation. There is no period after the abbreviation.
micron (10^{-6} meter)	μm	
millimeter (10^{-3} meter)	mm	
centimeter (10^{-2} meter)	cm	
meter	m	450 nm, 10 μm, 2.5 cm
MASS		
nanogram (10^{-9} gram)	ng	*Note*: There is a space between the number and the abbreviation. There is no period after the abbreviation.
microgram (10^{-6} gram)	μg	
milligram (10^{-3} gram)	mg	
gram	g	
kilogram (10^{3} gram)	kg	450 ng, 100 μg, 2.5 g, 10 kg

(Continued)

TABLE 7.5 (*continued*)

Term	Symbol or Abbreviation	Example
VOLUME		
microliter (10^{-6} liter)	μl or μL	*Note*: There is a space between the number and the abbreviation. There is no period after the abbreviation.
milliliter (10^{-3} liter)	ml or mL	
liter	l or L	
cubic centimeter (ca. 1 mL)	cm³	450 μl or 450 μL, 0.45 ml or 0.45 mL, 2 l or 2 L
TIME		
seconds	s or sec	*Note*: There is a space between the number and the abbreviation. There is no period after the abbreviation.
minutes	min	
hours	h or hr	
days	d	60 s or 60 sec, 60 min, 24 h or 24 hr, 1 d
CONCENTRATION		
molar (U.S. use)	M	TBS contains 0.01 M Tris-HCl, pH 7.4 and 0.15 M NaCl.
molar (SI units)	mol L⁻¹	
parts per thousand	ppt	Brine shrimp can be raised in 35 ppt seawater.
OTHER		
degree(s) Celsius	°C	15 °C (there is a space between number and symbol)
degree(s) Fahrenheit	°F	59 °F (there is a space between number and symbol)
diameter	diam.	pipe diam. was 10 cm
figure, figures	Fig., Figs.	As shown by Fig. 1, ...
foot-candle	fc or ft-c	500 fc or 500 ft-c
maximum	max	The max enzyme activity was found at 36 °C.
minimum	min	The min temperature of hatching was 12 °C.
mole	mol	
percent	%	95% (there is no space between number and symbol)
species (sing.)	sp.	Tetrahymena sp.
species (plur.)	spp.	Tetrahymena spp.

Numbers

Numbers are used for quantitative measurements. In the past, numbers less than 10 were spelled out, and larger numbers were written as numerals. The modern scientific number style recommended in the CSE Manual (2014) aims for a more consistent usage of numbers. The rules are as follows:

1. Use numerals to express any *quantity*. This form increases their visibility in scientific writing, and emphasizes their importance.
   - Cardinal numbers, for example, 3 observations, 5 samples, 2 times
   - In conjunction with a unit, for example, 5 g, 0.5 mm, 37 °C, 50%, 1 hr. Pay attention to spacing, capitalization, and punctuation (see Table 7.5).
   - Mathematical relationships, for example, 1:5 dilution, 1000× magnification, 10-fold

2. Spell out numbers in the following cases:
   - When the number begins a sentence, for example, *"Twelve g of turnips was* (not *were*) *homogenized."* rather than *"12 g of turnips was homogenized."* Alternatively, restructure the sentence so that the number does not begin the sentence. Notice that when numbers are used in conjunction with units, the quantity is considered to be singular, not plural.
   - When there are two adjacent numbers, retain the numeral that goes with the unit, and spell out the other one. An example of this is *The solution was divided into four 250-mL flasks.*
   - When the number is used in a nonquantitative sense, for example, *one of the treatments, the expression approaches zero, one must consider…*
   - When the number is an ordinal number less than 10, and when the number expresses rank rather than quantity, for example, *the second time, was first discovered.*
   - When the number is a fraction used in running text, for example, *one-half of the homogenate, nearly three-quarters of the plants.* When the precise value of a fraction is required, however, use the decimal form, for example, *0.5 L* rather than *one-half liter.*

3. Use scientific notation for very large or very small numbers. For the number 5,000,000, write 5×10^6, not *5 million*. For the number *0.000005*, write 5×10^{-6}.

4. For decimal numbers less than one, always mark the ones column with a zero. For example, write *0.05*, not *.05*.

TABLE 7.6	Checklist for proofreading format
Category	**Check for**
Section headings	Correct order, consistent format, not separated from section body
Lists (bulleted or numbered)	Sequential numbering and consistent style, parallelism in sentence structure, consistent indentation for each level
Figures and tables	Sequential numbering in the order they are described
In-text references to figure and table numbers	Correspondence with the actual figures and tables
In-text references	One-to-one correspondence with the end references; formatting is correct
End references	One-to-one correspondence with the in-text references; all information is present; formatting is correct
Headers and footers (if needed)	Correct position on each page
Page numbers	Sequential (check especially after section breaks in Microsoft Word)
Typography	Consistent typeface, font size, and line spacing

Format

Most university writing centers and professional editors recommend proofreading your paper in multiple "passes," looking for one kind of error in each pass (The Writing Center at UNC Chapel Hill 2020; CUNY Writing Fellows 2020; Every 2020). This strategy works particularly well for finding formatting errors, which are much easier to detect on printed pages than on the computer screen (Table 7.6). Check for potential errors in the following areas:

- Section headings
- Bulleted or numbered lists
- Figures and tables, including their in-text references
- References to outside sources within the text (in-text references)
- Full references at the end of the paper
- Headers and footers
- Page numbers
- Typography

7.4 Get Feedback

When we are engrossed in our work, we may fail to recognize that what is obvious to us is not obvious to an "outsider." That is where feedback from someone who is familiar with the subject matter comes in handy. If your instructor allows it, ask your lab partner, another classmate, or your teaching assistant to review your paper. Return the favor by reviewing someone else's. You may also get valuable feedback from a writing expert at your school's writing center.

The questions your reviewer will focus on are as follows:

- Do I know what the writer is trying to accomplish with this paper? Is the purpose clear?
- What questions or concerns do I have about this paper? Are there sections that were difficult to follow? Are the organization, content, flow, and level appropriate for the intended audience?
- What suggestions can I offer the writer to help him/her clarify the intended meaning?
- What do I like about the paper? What are its strengths?

Tips for being a good peer reviewer

There are two issues with which you may struggle when you are asked to review your classmate's paper: (1) I'm not confident that I know the "right" answer or know enough about the writing process to give good suggestions, and (2) I don't want to hurt the writer's feelings. These are valid concerns, and resolving them will require, first, a willingness to learn as much about writing scientific papers as possible, and second, the attitude that if something is unclear to you, it may also be unclear to other readers. With each paper you review, you will gain more confidence in your ability to give constructive feedback. In the meantime, however, a good rule of thumb is to give the kinds of suggestions and consideration that you would like to receive on your own paper.

When reviewing electronic files, the **New Comment** and **Track Changes** commands on the **Review** tab in Microsoft Word are very useful (see Section A1.6 in Appendix 1). **New Comment** allows the reviewer to make a comment or query, without editing the text itself, off to the side of the main text. When **Track Changes** is turned on, the reviewer's suggested changes, typed right into the body of the text, appear in a different color. The author can then accept or reject the changes. Think of the peer review process as a team sport: the reviewer is not challenging the writer's right to be on the team. The two are working together to get the best possible result.

Here are some concrete tips for being a good peer reviewer:

- Talk to the writer about his or her objectives, questions and concerns, parts that need specific feedback, and perceived strengths and weaknesses.
- Use the "Scientific Writing Checklist" for content.

- Look over Table 8.2 and "A Lab Report in Need of Revision" (Chapter 8) for common errors.
- Mark awkward sentences, spelling and punctuation mistakes, and formatting errors. Do not feel you have to rewrite individual sentences—that is the writer's job.
- Ask questions. Let the writer know where you can't follow his/her thinking, where you need more examples, where you expect more detailed analysis, and so on.
- Do not be embarrassed about making lots of comments; the author does not have to accept your suggestions. On the other hand, if you say only good things about the paper, how will the writer know whether the paper is accomplishing the desired objectives?

You can fine-tune your proofreading skills on any text. You may recognize some of your own problems in other people's writing, and, with persistence and practice, you will find creative solutions to correct these problems. **Keep a log of the problems that recur in your writing** and review them from time to time. Repetition builds awareness, which will help you achieve greater clarity in your writing.

Have an informal discussion with your peer reviewer

Sometimes the comments made by the peer reviewer are self-explanatory. Other times, however, the peer reviewer cannot respond to certain parts of the paper, because more information is required. Under these circumstances, an informal discussion between the writer and the reviewer is helpful. There are two important rules for this discussion:

- First, the writer talks and the reviewer listens. The objective is to help the writer express exactly what he/she wants to say in the paper.
- Second, the reviewer talks, in nonjudgmental terms, about which parts of the paper were readily understandable and which parts were confusing. When you are the reviewer, be respectful and considerate. Critique the writing, not the writer. Suggest, rather than prescribe, changes. Try to make suggestions specific to give the writer a sense of direction. Recommend authoritative sources.

Feedback from your instructor

Some instructors write comments on the hard copy of an assignment; others use electronic editing tools. When instructors grade hard copies, they usually use standard proofreading marks to save time. If you don't know what the marks mean, ask your instructor or look them up. Frequently used proofreading marks are listed in Table 8.1. A more comprehensive list is available in the CSE Manual (2014), at http://www.biomedicaleditor.com/support-files/proofreadingmarks.pdf, and in other printed and on-line sources.

Online submissions have become increasingly widespread, especially at the college level, for several reasons. Students like the convenience. Sub-mitting assignments online does not require a printer and can be done at the last minute. When a document is stored in the Cloud, it can be accessed from a variety of devices, and there is also less chance that it will get lost (physically, at least). Going paperless is also good for the environment.

Online submissions have benefits for instructors as well, which include:

- Being able to provide higher quality, more consistent feedback in less time.
- Having a time stamp to confirm when a student's assignment was turned in.
- Having the capability to check for plagiarism automatically.

Most of the concerns that students and instructors voice about online submissions are related to technical difficulties during the upload process, changes in document format, the inability to resubmit a document if an error is discovered after the fact, unreliable internet connections, and a personal dislike of technology in general. Ultimately, it is up to your in-structor to decide whether the advantages of online submission outweigh the disadvantages.

Instructors have various options for providing feedback in your elec-tronic documents. One option is to insert comments or mark up text us-ing the Comments and/or Track Changes features in Word (see Section A1.6 in Appendix 1). A second option that applies specifically to PDFs is to open the PDF in Adobe Acrobat Reader. In this program, text can be highlighted and comments can be added on "sticky notes." A third op-tion for providing feedback can be used on either Word documents or PDFs uploaded to Turnitin. In Feedback Studio, instructors can insert QuickMarks, comments, or text (Figure 7.1, item ①). **QuickMarks** are frequently-made comments that can be saved and reused on the cur-rent paper or on other students' papers (Figure 7.1, item ②). The title of a QuickMark is limited to 40 characters, but there is plenty of room to add detailed explanations in the context box. The next time this comment is applicable, instructors simply drag the QuickMark off the side panel and drop it onto the student's paper (Figure 7.1, item ③). Additional com-ments can be added to existing QuickMarks; these additional comments are only displayed in that particular comment in that particular paper (Figure 7.1, item ④). However, universal edits to existing QuickMarks will be applied to all papers in which that QuickMark was used. Two other features that instructors will appreciate are

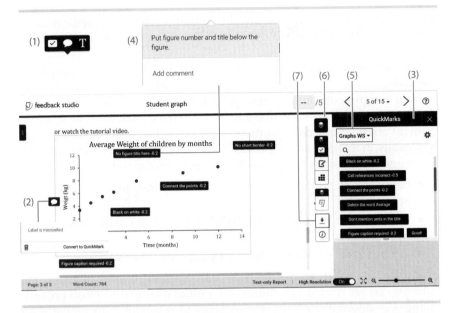

Figure 7.1 Comments and QuickMarks in Turnitin make it possible to provide consistent and detailed feedback on assignments. (1) Feedback can be made with QuickMarks, in comment balloons, or in text boxes. (2) Comments apply only to the current text, but comments saved as QuickMarks can be reused for all assignments. (3) QuickMarks in the right panel can be dragged and dropped onto the text. (4) Comments expand when the comment is clicked. (5) QuickMarks are organized into sets that can be customized for individual assignments. (6) In the student view, there is a blue comment balloon icon that students click to display their instructor's comments. (7) Assignments with comments can be downloaded as PDFs or printed for future reference.

- QuickMarks can be saved in sets customized for individual assignments (Figure 7.1, item ⑤) and
- QuickMark sets can be shared, which allows grading to be more consistent across multiple lab sections.

Students are able to see their instructor's comments by clicking the assignment link to open their paper, clicking the blue comment balloon on their list of icons (Figure 7.1, item ⑥), and then clicking on each comment (Figure 7.1, item ④). The document along with all of the expanded comments can be downloaded as a PDF or printed out for future reference (Figure 7.1, item ⑦).

7.5 Scientific Writing Checklist

TITLE (Section 5.6)
❏ Descriptive and concise

AUTHORS
❏ Each author's first name is followed by his/her surname

ABSTRACT (Section 5.6)
❏ Contains an introduction (background and objectives)
❏ Contains brief description of methods
❏ Contains results
❏ Contains conclusions

INTRODUCTION (Section 5.5)
❏ Starts with a general introduction to the topic
❏ Contains a question or unresolved problem
❏ Contains background information supported by in-text references
❏ The selected references are directly relevant to the study.
❏ The in-text reference is formatted correctly according to the Name-Year, Citation-Sequence, or Citation-Name system (Section 4.3).
❏ Information obtained from a reference is paraphrased. Direct quotations are not used (Sections 4.2 and 4.3).
❏ The objectives of the study are clearly stated.

MATERIALS AND METHODS (Section 5.3)
❏ Contains all relevant information to enable a person with appropriate training to repeat the procedure.
❏ Routine procedures are not explained.
❏ Complete sentences and paragraphs are used—do not make a numbered list.
❏ Past tense is used because these actions were done in the past and completed.
❏ Passive voice is used to emphasize the action (active voice is allowed in some disciplines).
❏ Materials are not listed separately.
❏ No preview is given of how the data will be graphed or tabulated.

RESULTS (Section 5.4)

❑ Tables and figures are described in numerical order. The descriptive text for a table or figure immediately precedes that table or figure (see, for example, Figures 5.1, 5.3, and 5.4).

❑ Results are described in *past* tense.

❑ Every sentence in the text is meaningful. Sentences such as *The results are shown in the figure below* are not meaningful.

❑ When a result is described, the figure showing that result is referenced, preferably in parentheses at the end of the sentence.

❑ There are no tables and figures that are not described.

❑ The figure caption is positioned *below* the figure. The table caption is positioned *above* the table.

❑ Figure and table titles are informative and can be understood apart from the text.

❑ The results are not explained.

DISCUSSION (Section 5.5)

❑ Each result in turn is *briefly* restated and then explained or interpreted.

❑ Past tense is used when referring to your own results. Present tense is used to state scientific fact, which is information supported by experimental evidence and replicated by many different scientists. Results in journal articles are considered to be fact until other studies present evidence to the contrary.

❑ The results are compared with those in journal articles.

❑ The results are related back to the original objectives stated in the Introduction.

❑ Any errors and inconsistencies may be pointed out.

❑ The significance of the results or their implications may be discussed in a broader context.

REFERENCES (Chapter 4)

❑ The references consist mostly of journal articles, not secondary sources such as textbooks or websites.

❑ The references are formatted correctly and contain all the required information.

❑ All references listed in this section have been cited in the text. All in-text references have been included in this section.

❑ Reference management software saves time formatting references.

REVISION

❏ All questions from the laboratory exercise have been answered.
❏ Calculations and statistics have been double-checked (Section 7.2).
❏ The overall structure of the manuscript is correct (Section 7.2).
❏ The overall structure of each section is correct (Section 7.2).
❏ Figures and tables are formatted correctly (Section 5.4).
❏ Sections, paragraphs, sentences, and words are coherent and meaningful (Section 7.2).
❏ Individual words are used appropriately for the situation (Section 7.2).
❏ All sentences are grammatically correct (Section 7.3).
❏ All words are spelled correctly (Section 7.3).
❏ The correct punctuation marks are used (Section 7.3).
❏ Abbreviations for unfamiliar terms are defined the first time they are used (Section 7.3).
❏ Standard abbreviations are used for units (Table 7.5).
❏ Numbers are formatted correctly and, when applicable, are followed by units (Section 7.3).
❏ The format for section headings, lists, figures, tables, references, and typography is consistent (Table 7.6).

Check your Understanding

1. Apply the revision strategies described in this chapter to a scientific paper (yours or a classmate's). Edit the big things first:

   • Overall structure: Are all of the sections present and are they in the right order?

   • Within each section (except the abstract), is the text divided into paragraphs? Does each paragraph focus on just one topic?

   • Within each section, is there a logical order to the paragraphs? The Introduction typically flows from general to specific. The procedure is described sequentially in the Materials and Methods. The Discussion flows from specific to general.

   • Within each paragraph, can you see the relationship between one sentence and the next? Look for signal words and phrases (transitions).

   • Does each sentence say something meaningful? Remove repetitive sentences, and replace them with a concise alternative.

   • Check that each sentence is technically accurate; this includes the description of the trends in the figures.

- Is it clear what the subject is doing in each sentence? Active voice makes sentences more dynamic, but in the Materials and Methods section, passive voice is preferred. Emphasize what was done, not that you did it.
- Pay attention to tense. Past and present tense have specific connotations in scientific writing.
- Pay attention to words. Is this the right word for the context? Watch for easily confused word pairs like *effect* and *affect*.
- Can fewer words express the same idea? Using more words than necessary doesn't make you sound smarter. It makes your readers work harder than they should.
- Make sure the noun that words like *this*, *that*, and *its* are referring to is unambiguous.
- Check that unfamiliar jargon and all acronyms are defined upon first mention.
- Check that the language is formal and gender-neutral.
- Check the layout of the figures. Does each one have a purpose? Are the figures arranged in a logical order?
- In the Results section, are the important trends in each figure described, and each figure is referenced by number? Does the text describing each figure immediately precede the figure?
- In the Discussion section, is each result summarized and interpreted?

    After editing the big things, look for grammatical, spelling, and punctuation errors.

    Finally, check each of the following elements for potential formatting errors. Scan the document specifically for one of these elements at a time (one element per "pass"):

- Are the section headings in the correct order and, if numbered, sequential?
- Is the style of the items in bulleted or numbered lists consistent?
- Are the figures and tables numbered sequentially? Do the in-text references correspond to the correct figure or table?
- Is each in-text reference to outside sources included in the references list at the end?
- Does each full reference have an in-text counterpart?
- Is the reference style consistent throughout the paper?
- If there are headers and footers, do they display correctly on each page?
- Are the page numbers displayed sequentially?
- Is the typeface, font size, and line spacing consistent throughout the paper?

2. Answer these true-false questions about the peer review process.
   - Peer review is getting feedback from a colleague who is familiar with the topic.
   - Peer review is only done by career scientists and professors.
   - Your writing will only be evaluated when you are in college. After that, you are on your own.
   - The main purpose of peer review is to proofread the paper for spelling and grammatical errors.
   - When making comments and suggestions, a good rule of thumb is to provide constructive feedback in a respectful tone.

3. Decide which of the comments in each pair would be more helpful to you as an author. Explain why.

   Example A

   (1) This is confusing.

   (2) I got confused here.

   Example B

   (1) This calculation procedure makes no sense.

   (2) There might be a mistake here. Consider rereading the lab handout. Consider rechecking the formulas in Excel.

   Example C

   (1) Wordy.

   (2) There are some good ideas here, but try to state them more concisely. Consider removing repetitive sentences.

   Example D

   (1) Everything looks good.

   (2) The structure and organization are what I expected.

4. It is natural to be disappointed when you receive negative feedback on your work. Decide which response in each pair would help you move forward and help you become a better writer.

   Example A

   (1) My peer reviewer put time and effort into looking over my paper.

   (2) My peer reviewer just wanted to make me feel inferior with all these comments.

Example B

(1) I'm so happy that my peer reviewer had only positive things to say. There is nothing I need to do to improve this paper.

(2) I'm a little concerned that my peer reviewer may have missed some things. I better make sure I've followed all the instructions before turning in my paper.

Example C

(1) I don't agree with my peer reviewer's suggestion.

(2) I better double-check the instructions in my lab handout (or look up this topic in my textbook or ask my professor). Maybe I overlooked something.

Example D

(1) I don't understand my peer reviewer's comment, so I'm just going to ignore it. It probably wasn't important.

(2) I think my peer reviewer may have misunderstood what I meant to say. I'm going to ask them to clarify their comment.

5. Peer review a classmate's paper. Apply the tips described in Section 7.4 for being a good peer reviewer.

Go to **macmillanlearning.com/knisely6e** and select "Student Site" to access samples, template files, and tutorial videos

Sample Student Laboratory Reports

Objectives

8.1 Contrast the features of a "good" student lab report with those of a lab report in need of improvement

8.2 Understand common proofreading marks and comments used by instructors to provide feedback

8.1 A "Good" Sample Student Lab Report

The first laboratory report in this chapter was written by Lynne Waldman during her first year at Bucknell University, in an introductory course for biology majors. Lynne and her lab partners designed and carried out an original project in which they investigated the effect of a fungus on the growth of bean, pea, and corn plants.

Lynne's report has many of the characteristics of a well-written scientific paper. When you look over her presentation, notice the style and tone of her writing, as well as the format of the paper. The comments and annotations in the margins alert you to important points to keep in mind when you write your laboratory report.

The presentation here has been typeset to fit this book and to accommodate the annotations. Your report should be formatted to fit standard 8.5 × 11 inch paper. Unless you are instructed otherwise, use a serif type (Times Roman is standard), double space, and leave at least 1 inch of margin all around.

For details on how to format documents in Microsoft Word, see Section A1.8 in Appendix 1.

## Chapter 8

The Effects of the Fungus *Phytophthora infestans* on Bean, Pea, and Corn Plants

Lynne Waldman, Partner One, Partner Two

Title is informative.

Write author's name first followed by lab partners' names.

Label sections of lab report clearly.

Provide background information.

State purpose of current experiment.

Briefly describe methods.

Do not cite sources in abstract.

Do not refer to any figures.

Describe results.

Briefly explain the results or state your conclusions.

Limit abstract to a maximum of 250 words.

Italicize Latin names.

Provide background information.

Abstract

Phytophthora infestans is a fast-spreading, parasitic fungus that caused the infamous potato blight by devastating Ireland's crops in the 1840s. *P. infestans* also causes late blight in tomato plants, a relative of the potato. In this experiment, the effects of *P. infestans* on *Phaseolus* variety long bush bean, *Zea mays* (corn), and *Pisum sativum* (pea) were studied. The soil surrounding the roots of 18-day old plants was injected with *P. infestans* cultured in an L-broth medium. Plant height, number of leaves, and leaf angle were measured for each plant during the next 8 days. Chlorophyll assays were performed prior to exposure, and on the eighth day after exposure to the fungus. The plants were also examined for black or brown leaf spots characteristic of late blight infections. The results showed that *P. infestans* had no apparent effect on the bean, corn, and pea plants. One reason for this may be that there were no fungus zoospores in the L-broth medium. More probably, however, *P. infestans* may be a species-specific pathogen that cannot infect bean, corn, or pea plants.

Introduction

Originating in Peruvian-Bolivian Andes, the potato (*Solanum tuberosum*) is one of the world's four most important food crops (along with wheat, rice, and corn). Cultivation of potatoes began in South America over 1,800 years ago, and through the Spanish conquistadors, the tuber was introduced into Europe in the second half of the 1600s. By the beginning of the 18th century, the potato was widely grown in Ireland, and the country's economy heavily relied on the potato crop. In the middle of the 19th century, Ireland's potato crop suffered widespread late blight disease caused by *Phytophthora infestans*, a species of pathogenic plant fungus. Failure of the potato crop because of late blight resulted in the Irish potato famine. The famine led to widespread starvation and the death of about a million Irish people.

The potato continues to be one of the world's main food crops. However, *P. infestans* has reemerged in a chemical-resistant form in the United States, Canada, Mexico, and Europe (McElreath, 1994). Late blight caused by the new strains is costing growers worldwide about $3 billion annually. The need to apply chemical fungicides eight to ten times a season further increases the cost to the grower (Stanley, 1994 and Stanley, 1997). *P. infestans* is thus an economically important pathogen.

P. infestans, which can destroy a potato crop in the field or in storage, thrives in warm, damp weather. The parasitic fungus causes black or purple lesions on a potato plant's stem and leaves. As a result of infection by this fungus, the plant is unable to photosynthesize, develops a slimy rot, and dies. *P. infestans* similarly infects the tomato plant (*Lycopersicon esculentum*) (Brave New Potato, 1994).

The purpose of the present experiment was to determine the effects of *P. infestans* on plant height, number of leaves, leaf angle, and chlorophyll content of three agriculturally important plants: *Phaseolus* variety long bush beans, *Zea mays* (corn), and *Pisum sativum* (peas). Symptoms of fungal infection were assumed to be similar to that in potatoes.

Materials and Methods

Phaseolus variety long bush bean, *Zea mays* (corn), and *Pisum sativum* (pea) seeds were soaked overnight in tap water. Fifteen randomly chosen seeds of each species were planted 1 cm beneath the surface in three separate trays containing 10 cm of potting soil. Another set of trays, which was to be the control group, was prepared in the same fashion. All the experimental plants were placed in one fume hood, and all the control plants were placed in relative positions in another fume hood in the same room. The plants were exposed to the ambient light intensity in the hood (153 fc) and air current 24 hrs a day, and

Margin notes:

Use proper citation format (e.g., Name-Year system).

Do not use direct quotations. Paraphrase source text and cite the source in parentheses.

Use an abbreviated title when no author is given.

State purpose of experiment clearly.

Write Materials and Methods section in past tense.

Provide sufficient detail to allow the reader to repeat the experiment.

were watered lightly daily. The plants were allowed to germinate and grow for 18 days.

Phytophthora infestans on potato dextrose agar was obtained from Carolina Biological Supply House. At day 10 of the plant growth regime, pieces of agar on which the fungus was growing were transferred to L-broth. L-broth consisted of 5 g yeast extract, 10 g tryptone, 1 g dextrose, and 10 g NaCl dissolved in distilled water, and adjusted to pH 7.1, to make 1 L of medium. The medium was sterilized before adding the fungal culture. After 4 days in L-broth, 6 mL of the fungal culture was injected into the soil around the roots of each 18-day old plant. Six mL of L-broth without *P. infestans* was injected into the soil of the control plants. All plants were then allowed to grow for another 8 days.

Every other day after treatment with *P. infestans*, plant height and number of leaves were measured for both the control and the experimental plants. Plant height was measured from the soil to the apical meristem of the plant. Leaf angle (as shown in Figure 1) of the largest, lowest leaf on each plant was measured three times, once prior to injection, once 4 days after injection, and once 8 days after injection. Leaf angle was measured in order to determine if *P. infestans* causes wilting in the three plant species. In addition, the plant was examined visually for the presence of any leaf spots.

Chlorophyll assays were performed on one plant from each tray prior to injection and on the eighth day after injection. For each chlorophyll assay, the leaves of the plant were removed from the stem. For each 0.1 g of

Include figures in Materials and Methods section if they help clarify the methodology.

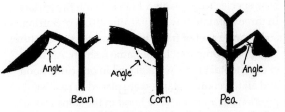

Figure 1   Leaf angle as measured in bean, corn, and pea plants

leaves, 6.0 mL of 100% methanol were used. The leaves were thoroughly ground in half of the methanol with a pestle in a mortar. The leaves were ground again after the rest of the methanol was added. Extraction of the chlorophyll was allowed to proceed for 45 min at room temperature. Then the suspension was gravity filtered through filter paper to remove the leaf parts. The absorbance of the filtrate was measured with a Spectronic 20 spectrophotometer at 652 nm and 665.2 nm. The absorbance values were converted to relative chlorophyll units using the following equation derived by Porra and colleagues (1989):

Total chlorophyll (a and b) = Dilution factor × [22.12 $A_{652\,nm}$ + 2.71$A_{665.2\,nm}$ (mg/L)] × Volume of solvent (L) / Weight of leaves (mg)

Make proper subscripts.

Results

P. infestans-treated plants and the control plants had similar growth patterns (Figure 2). Both the experimental and control pea and corn plants grew at a constant, but very slow rate over the eight day test period. The control bean plants were taller on average than the experimental bean plants throughout most of the experiment. Both groups showed the same growth pattern, however, with rapid growth occurring from day 18 to 24 (0 to 4 days after injection), followed by slower growth to the end of the experiment.

As plant height increased, the average number of leaves on all of the plants also increased over the measurement period (Figure 3). There is an

Include text in the Results section. Describe the important results shown in each figure and table.

Refer to each figure and table in parentheses.

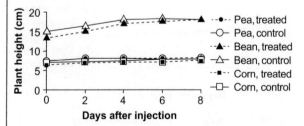

Make text in legend and in axes titles large enough to read easily.

Make sure intervals on axes have correct spacing.

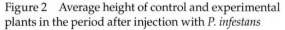

Figure 2 Average height of control and experimental plants in the period after injection with *P. infestans*

Make points and lines black and background white for best contrast.

In the axes titles, write the variable followed by the units in parentheses (where applicable).

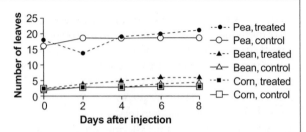

Position the figure caption below the figure.

Figure 3   Average number of leaves of control and experimental plants in the period after injection with *P. infestans*

uncharacteristic decrease in the number of leaves of pea plants treated with *P. infestans* from day 18 to 20 (0 to 2 days after injection), but this is probably due to counting error.

There was a general decline in average leaf angle of all the plants over the first four days after injection with *P. infestans* (Figure 4). The plants did not follow this pattern over the second half of the experiment, however. The leaf angle of the experimental bean group increased by 28°, while that of the control bean group only increased by about 3°. The leaf angle of the control pea plants increased significantly (33°), while that of the experimental pea plants decreased 4°. The leaf angle of the corn control group decreased 0.5°, while that of the corn experimental group showed a much sharper decline of 24°.

There was also no difference between the experimental and control groups with regard to

Describe the figures in order.

Insert symbols such as ° using word processing software.

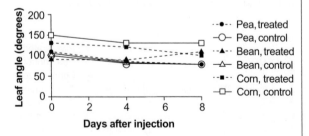

Make the figure title descriptive. Do not use "y-axis title vs. x-axis title."

Figure 4   Average leaf angle of control and experimental plants in the period after injection with *P. infestans*

Table 1   Chlorophyll content of corn, bean, and pea plants prior to infection and 8 days after infection

Position the table caption above the table.

Plants	Relative chlorophyll units		Change in chlorophyll content (relative units)
	Day 0	Day 8	
Corn, treated	9.036×10^{-4}	9.383×10^{-4}	$+3.45 \times 10^{-5}$
Corn, control	9.270×10^{-4}	8.963×10^{-4}	$+3.34 \times 10^{-5}$
Bean, treated	1.034×10^{-1}	1.2×10^{-3}	-1.022×10^{-1}
Bean, control	1.7×10^{-3}	1.6×10^{-3}	-1×10^{-4}
Pea, treated	1.3×10^{-3}	1.7×10^{-3}	$+4 \times 10^{-4}$
Pea, control	1.2×10^{-3}	1.2×10^{-3}	0.0000

Use scientific notation correctly.

Do not split small tables across two pages.

chlorophyll content. There was a slight increase in chlorophyll content from day 18 to 26 (0 to 8 days after injection) in the corn plants (Table 1). For the bean group, there was a large decrease in chlorophyll content, 0.1 relative chlorophyll units, which did not seem to agree with the general appearance of the plants. There may have been some error when this assay was carried out. There was little change in chlorophyll content for the pea group.

Number tables independently of figures.

Finally, there was no evidence of any brown or black leaf spots symptomatic of *P. infestans* infection.

Discussion

P. infestans did not affect the plant height, leaf angle, number of leaves, and chlorophyll content of *Zea mays*, *Pisum sativum*, or *Phaseolus*. Symptoms of infection are the presence of brown or black spots (areas of dead tissue) on leaves and stems, and, as the infection spreads, the entire plant becomes covered with a cottony film (Stanley, 1994). None of the experimental plants exhibited these symptoms.

Briefly restate the results in the Discussion section.

There may be several reasons why *P. infestans* did not affect the plants in this study. One reason is that the L-broth culture of *P. infestans* may not have contained zoospores of the fungus. Zoospores are motile spores that can penetrate the host plant through the leaves and soft shoots, or through the roots (Stanley, 1994). Zoospores are usually produced in wet,

Give possible explanations for the results.

Support explanations with references to published sources.

warm weather conditions (Ingold and Hudson, 1993). If the L-broth culture did not contain any zoospores, or if the soil around the plants was not sufficiently saturated to stimulate production of zoospores, then these conditions may have prevented *P. infestans* from attacking the roots and shoots of the plants.

In order to determine if the problem was lack of zoospores, first the L-broth culture could be examined microscopically for presence of zoospores. Second, the *P. infestans* plants could be watered with different quantities of water to determine if the fungus requires wetter soil for zoospore production and motility.

Another reason why *P. infestans* may not have affected the plants is that this species of fungus may be specific to potato (*Solanum tuberosum*) and tomato (*Lycopersicon esculentum*) plants (Stanley, 1994), which both belong to the nightshade family (Solanaceae). In contrast, corn belongs to the grass family (Gramineae), and peas and beans are legumes (Leguminosae). It may be that these plant families are not susceptible to *P. infestans*, which has a very limited host range (Stanley, 1994). Non-susceptible plants have been shown to have defense mechanisms that prevent *P. infestans* from infecting them (Gallegly, 1995).

Further research is required to determine if *P. infestans* really cannot infect corn, pea, and bean plants. Goth and Keane (1997) developed a test to measure resistance of potato and tomato varieties to original and new strains of *P. infestans*. Their experiments involved exposing the experimental plants' leaves directly to the fungus, and this method could perhaps be tested on corn, pea, and bean leaves as well.

References

Brave New Potato. 1994. Discover 15(10): 18–20.

Gallegly ME. 1995. New criteria for classifying Phytophthora and critique of existing approaches. In: Erwin DC, Bartnicki-Garcia S, Tsao PH, editors. Phytophthora: Its Biology, Taxonomy, Ecology, and Pathology St. Paul: The American Phytopathological Society. pp. 167–172.

Margin notes:

Offer possible ways to test whether explanation is valid.

Whenever possible, use primary references (journal articles, conference proceedings, collections of primary articles in a book). Avoid unreliable websites.

Substitute the title when no author is given.

In Name-Year end reference format, list authors alphabetically by first author's last name.

Goth RW, Keane J. 1997. A detached-leaf method to evaluate late blight resistance in potato and tomato. American Potato Journal 74(5): 347–352.

Ingold CT, Hudson HJ. 1993. The Biology of Fungi, 6th ed. London: Chapman and Hall.

McElreath, Linda R. 1994. One potato, two potato. Agricultural Research 42(5): 2–3.

Porra RJ, Thompson WA, Kriedemann PE. 1989. Determination of accurate extinction coefficients and simultaneous equations for assaying chlorophylls a and b extracted with four different solvents: verification of the concentration of chlorophyll standards by atomic absorption spectroscopy. Biochimica et Biophysica Acta 975: 384–394.

Stanley D. 1994. What was around comes around. Agricultural Research 42(5): 4–8.

Stanley D. 1997. Potatoes. Agricultural Research 45(5): 10–14.

Use mostly primary journal articles or articles in a book.

List all authors (up to 10; then list first 10 followed by *et al.* or *and others*).

See the tabbed pages in Chapter 4 for examples of how to reference printed and electronic sources.

Give inclusive page numbers, not just the page(s) you extracted information from.

Make sure all in-text citations have a corresponding end reference.

Make sure all end references have a corresponding in-text citation.

8.2 Laboratory Report Errors

Table 8.1 lists proofreading marks that your instructor may use to give you feedback on your lab report. Marks are placed in the margin as well as in the text itself to describe the required revision.

Table 8.2 lists some common errors made by students writing biology lab reports. As instructors, we want students to be aware of these and other errors and to understand how to correct them. However, providing detailed explanations takes time, and time is a precious commodity. To save time without compromising quality, some instructors use tools for editing electronic documents, such as Word's Comments and Track Changes features and Turnitin's QuickMarks. Other instructors prefer to write abbreviated comments in the margin of printed documents. Either way, as a student, it is your responsibility to take your instructor's comments seriously and take the necessary steps to avoid making the same kinds of errors in the future. Tables 8.1 and 8.2 can be downloaded from **macmillanlearning.com/knisely6e**. Instructors can edit the tables to customize the abbreviations for specific assignments. Students are expected to look up the meaning of each abbreviated comment for continuous improvement of their writing skills.

Lab Report Errors

TABLE 8.1	A short list of proofreading marks	
In Margin	**In Text**	**Explanation**
[	[More starch	Left-align text
]	(1)]	Right-align text
][	]Lab report title[	Center text
lc	PH	Use lowercase letter
CAP or uc	dna	Use capital (uppercase) letter
^to	refer ^ each	Insert text
ℓ	absolutely essential	Delete text
#	in|spite of	Insert space
⌒	to morrow	Close up
¶	starch. ¶The objective	Start a new paragraph
......	at room temperature	A dotted underline means "stet," or "let original text stand." The correction was made in error.
ital	E. coli	Italicize Latin names of organisms
∨	6.02 × 10E23	Delete E and superscript exponent
∧	Vmax	Subscript needed

TABLE 8.2	Abbreviations for comments made on lab reports
Abbreviation	**Explanation**
abbr	Write out the abbreviation the first time it is used, e.g., wild type (WT).
agr	Make subject and verb agree.
awk/incomplete	Revise to make less awkward. Convert sentence fragment to a complete sentence.
bkgd	Background is insufficient. Add details.
calc guid	Use full sentences to guide the reader through the calculation procedure. Use past tense (because the procedure was done in the past) and passive voice (to emphasize what was done, not who did it). For example, *The ___ was calculated using ___.* Show the original equation and define the terms. Then substitute known (or measured) values and solve for the unknown. State the final answer and include units.
caption pos	Figure caption goes *below* the figure. Table caption goes *above* the table. When you make the figure in Excel, leave the "Chart title" space blank.

TABLE 8.2	Abbreviations (*continued*)
Abbreviation	**Explanation**
cit form N-Y	• For 2 authors, include both authors' surnames separated by *and* followed by year of publication.
	• For 3 authors, include first author's surname followed by *and others* and year of publication.
cit form	Use CSE Name-Year, Citation-Sequence, or Citation-Name format. See Chapter 3 for specific examples. Do not use direct quotations. Paraphrase and cite the source.
cit missing	Cite all end references in the text.
command	Do not use command style. Reword in past tense. For example, rather than *Substitute the absorbance for y and solve for x*, write *The absorbance was substituted for y, and the equation was rearranged to solve for x*.
compare	• Compare treatment and control groups.
	• Compare your results with those in journal article.
content	Section is missing essential content
details	• The title is not descriptive. Add details such as variables and organism(s).
	• Essential details are missing in the Materials and Methods section. Provide enough detail to enable a trained person to repeat the experiment.
don't preview	In the Materials and Methods section, do not give a "preview" of how data will be plotted or tabulated in the Results section.
eq ed	Use Equation Editor in Word to make professional-looking equations. See Section A1.5 in Appendix 1.
expl	Explanation is insufficient.
fig format	Figure format is incorrect. See Section A2.4 in Appendix 2 to format graphs in Excel.
	• Use CSE-preferred symbols: filled or open circles, squares, and triangles.
	• Make all lines and symbols black for best contrast.
	• Use outside tick marks.
	• On the axis label, put units in parentheses after the variable.
	• Use standard intervals in multiples of 2 or 5 on the axes.
	• Shorten axis to eliminate empty space.
	• Legend is not needed when there is only one data set.

Lab Report Errors

TABLE 8.2 Abbreviations (continued)	
Abbreviation	**Explanation**
	• Legend is needed to distinguish multiple data sets on one graph.
	○ Move legend inside axes.
	○ Make legend entries meaningful. See "More than one data set" in Section A2.4 of Appendix 2.
	• Insert a line to show the trend. See "Choose a line" in Section A2.4 to decide which type of line.
	• If gridlines are used on bar graphs, make them unobtrusive.
	• Delete chart border.
fig/tab pos	Position the figure/table immediately after the text where it is first described. That way readers will read the description first and know what to expect when they see the data.
fig/tab ref	Reference figure/table number in parentheses at the end of the sentence. Put the period after the closing parenthesis.
fig/tab title	• Figure/table title is factually incorrect.
	• Figure/table title is inadequate. Add details to make title self-explanatory.
	• Use sentence case (do not capitalize common nouns unless they start a sentence).
gram	Grammatical error
head-body sep	Keep section heading and body together.
	• **Windows:** On the ribbon, **Home | Paragraph diagonal arrow | Line and Page Breaks** tab. Check the **Keep with next** checkbox.
	• **Mac:** On the menu bar, **Format | Paragraph | Line and Page Breaks**. Check the **Keep with next** checkbox.
heading	Add section heading.
hyp	State the hypothesis.
interp	Interpret the results in the Discussion (not the Results) section.
math	Calculation error
meaning	Make sentences meaningful.
not a recipe	Do not list materials separately. Do not make a numbered or bulleted list. Use full sentences, paragraphs, and past tense to describe the procedure.
num format	• Put a zero in the ones place: 0.1 mL not .1 mL
	• Use scientific notation when numbers are very large or very small.

TABLE 8.2	Abbreviations (*continued*)
Abbreviation	**Explanation**
obj?	State the objective(s) of the current study.
page break	End the current page; move subsequent text to next page. • **Windows: Ctrl+Enter** • **Mac: ⌘+Enter**
¶	Break this section into paragraphs. When you start a new topic, start a new paragraph.
passive voice	Passive voice is preferred. Shift the emphasis from yourself to the subject of the action.
past tense	• Use past tense to state the objectives. • Use past tense to describe the procedure. • Use past tense to indicate that you are referring to your own results and not making a statement that is universally true.
present tense	Use present tense for scientific fact (information already accepted by the scientific community).
punc	Punctuation error
ref missing	List all in-text references in the References section.
ref format	Reference elements are out of order. Information is missing.
rep	Eliminate repetition.
result?	• Describe the result, and reference the figure/table where the data are located. • Use specific language. How did the independent variable affect the dependent variable? What was the general relationship or trend?
round up	Round up final answer. The final answer cannot be more precise than the least precise measured value.
routine	Do not describe routine laboratory procedures in detail.
run-on	Break run-on sentence into two sentences.
scatter not line	Choose "XY Scatter" not "Line" in Excel to space data correctly. See Section A2.4 in Appendix 2.
source?	Cite sources to provide background information, to substantiate claims, and to compare findings.
sp	Word is misspelled
sub/super	• Superscript exponents. • Use AutoCorrect to format expressions with sub/superscripts automatically. See Section A1.10 in Appendix 1.

Lab Report Errors

TABLE 8.2	Abbreviations (*continued*)
Abbreviation	**Explanation**
symbol	Choose the correct symbol from **Insert	Symbol** on the Ribbon. For example, insert °C instead of writing out *degrees Celsius*. See Section A1.11 in Appendix 1 to define hot keys.
tab unnec	Do not include a table when the graph shows the same data.
units	Units are missing or incorrect.
unnec intro	Do not write unnecessary introductory phrases like this. See Section 5.4.
wc	Word choice. This word does not fit the context. See Section 7.2, "Choose your words carefully."
wordy	Revise to reduce wordiness.

8.3 A Lab Report in Need of Revision

The following lab report contains the kinds of errors that are made by students who are just learning to write biology lab reports. Some of these errors have to do with writing in general (punctuation, word usage, sentence structure, transitions, coherent paragraphs, and so on) and others are specific to writing scientific papers in biology. After writing but before revising the first draft of your lab report, skim the "lab report in need of revision" to anticipate the kinds of errors often made in the various sections. Use Tables 8.1 and 8.2 to look up the meaning of the abbreviations in the margin. Although some of these errors may not affect the content of your paper, they do affect your credibility as a scientist; careless writing may be equated with careless science.

Details: What aspect of barley seeds did you study?
Center title and author
Delete *by*
abbr: ∧gibberellic acid (GA)
/ wc: ∧affects / ∧include common name barley
ital. Latin names
Passive voice: use for procedure
Content: results and disc missing

]Barley seed lab report[

]by~~ Ima Sprout[

Abstract
We looked at how ∧ (GA) ~~effects~~ ∧ the production of α-amylase in ∧ Hordeum vulgare seeds. We cut seeds in half and gave them either water or a concentration of GA. After a week we measured how much starch was left in the seed. Our results showed that there was a correlation between GA concentration, how much starch was remaining, and the amount of α-amylase produced in the seeds.

Lab Report Errors

Introduction

(The concentration of GA was experimented on to see how this factor effected production of a-amylase.) When a barley seed germinates, (GA) is released from the embryo. GA moves to the aleurone layer of the seed, where it activates the production of α-amylase (Tanaka). After its release, α-amylase diffuses through the endosperm and hydrolyzes starch into α-1,4-linked glucose oligosaccharides that are then broken down by other enzymes into maltose and glucose (Koning). There should thus be a correlation between the amount of GA and the digestion of the endosperm in barley seeds.

Our hypothesis was that higher GA concentrations would cause increased α-amylase production, which would cause a decrease in starch in the barley seed endosperm.

Materials and Methods

~~Obtain filter paper, forceps, a razor blade, 3 Petri dishes, barley seeds, 10 μM GA, 10 mM HEPES-EGTA-Ca2+, a centrifuge, starch solution, HCl, 10 mM citric acid-sodium citrate buffer, Lugol's iodine, and a spectrophotometer. The tests had either~~ Wwhole barley seeds (Plate A) or half barley seeds (Plates B-F). ~~The half seeds had~~ ^ the embryo removed ~~and the~~ ^ cut ~~endosperm was placed~~ ^ down on the 3 pieces of filter paper in ~~the~~ ^ Petri dish. ~~Next, 1, .1, and .01 concentrations of GA were made using a serial dilution of 10 μM GA and 10 mM HEPES-EGTA-Ca2+.~~ Then 3.5 mL of ~~buffer~~ ^ or ^ hormone ~~were~~ ^ added to each of the Petri dishes according to the following table.

Margin notes:
- move obj to the end of the intro and make sentence less wordy; wc: affected not effected
- define abbr
- cit form N-Y
- cit form N-Y
- not a recipe
- ^with
- ^were placed
- ^side
- ^a
- ^10 mM HEPES-EGTA-Ca2+ buffer / 1, 0.1, and 0.01 μM GA / solution was
- num format (fix every time)

Plate	Seeds	Treatment
A	Whole seeds	3.5 mL buffer
B	Half seeds	3.5 mL buffer
C	Half seeds	3.5 mL (.01) μM GA
D	Half seeds	3.5 mL (.1) μM GA
E	Half seeds	3.5 mL 1 μM GA
F	Half seeds	3.5 mL 10 μM GA

Lab Report Errors

The Petri dishes were wrapped in foil and placed in a ~~15 degrees Celsius~~ ^ chamber for one week.

symbol: ^15 °C
ɣ

After 1 week ~~the plates were collected and~~ approximately 1 g of seeds ~~were~~ ^ put into a mortar ~~that was~~ on ice. The seeds were pulverized with 10 mL cold 10 mM citric acid-sodium citrate buffer at pH 5 for 5 min. The pulverized seeds were then placed in a centrifuge tube and the mortar was rinsed with 15 mL ice cold buffer to get the seed product into the tube. ~~This was done separately for each one of the plates.~~ The six samples were centrifuged for 10 min at 15,000G at ~~2 degrees Celsius~~ ^. ~~The supernatant was collected.~~ One mL of each ~~one of the six~~ supernatant was ~~added to six tubes at room temperature. Then in 20 second intervals~~ ^ 2 mL of .025% starch ~~was added to each tube successively.~~ After 120 sec, 7 mL of 1 N HCl was added to ~~the first tube and then down the line every 20 seconds. One~~ ^ mL of Lugol's iodine solution ~~was added to each tube~~. Lastly, the absorbance ~~value~~ of each ~~tube~~ ^ was taken with a Spec 20 spectrophotometer at 580 nm.

1 g singular: ^was

wordy
ɣ
abbr: ^2 °C

^mixed with

^each sample, followed by 1

wc: ^sample

The ~~experimental results of~~ starch ~~concentrations were compared to a standard curve. Using a 0.05% stock solution, a serial dilution was used to make~~ ^ .025%, .0125%, .006% and .003% ^ ~~concentrations. Two mL of each solution was poured into 5 different tubes. A sixth tube (blank) was filled with 2 mL water. Then~~ 1 mL buffer, 7 mL HCl, ~~annd~~ 1 mL Lugol's iodine solution ~~was added to each tube~~. The absorbance ~~value for each tube~~ was determined ~~by a Spec 20 spectrophotometer~~ at 580 nm. ~~Absorbance was plotted on the y-axis and percent starch was plotted on the x-axis. A best fit trend line was inserted to determine the equation of the standard curve.~~

^standards were prepared by combining 2 mL of / starch solution with

sp

wordy

don't preview

Results

~~The results from the experiment support the hypothesis that the digestive enzyme α-amylase production is dependent on the presence of gibberellic acid (GA) in germinating seeds. α-amylase promotes plant growth growth by digesting starch in the seed's endosperm. The higher the gibberellic acid content,~~

interp
ɣ

the higher the amount of α-amylase production and starch digestion.

The experiment included six groups of germinating seeds. Whole seeds in a 3.5 mL buffered solution served as the positive control because their embryos naturally produce gibberellic acid. They were expected to produce α-amylase and thus exhibit starch digestion. The negative control seeds were the endosperm halves of seeds in a 3.5 mL buffered solution because they had no way of synthesizing or obtaining gibberellic acid. The negative control group was not expected to produce any α-amylase or digest as much starch as the other groups. The other four groups of endosperms received treatments of 3.5 mL GA in varying concentrations.

The starch standard curve indicates a positive relationship between light absorbance at 580 nm and starch concentration. The equation of the standard curve (Figure 1) was used to calculate the percentage of starch remaining in the endosperm of the seeds after each treatment. The absorbance values determined for each of the six treatments were substituted into the equation for y to determine the remaining starch concentration (x).

unnec: absorbance is supposed to be proportional to concentration by Beer's Law

The whole seed positive control and the GA-treated half seeds all exhibited lower percentages of ~~absorbance~~ ^ than the negative control. The ~~absorbance level and~~ starch (%) decreased as the concentration of the 3.5mL μM gibberellic acid (GA) increased ~~per treatment. The whole seeds had 0.0216% of their starch remaining, while the 0.01 μM GA, 0.1 μM GA, 1 μM GA, and 10 μM GA had 0.0235%, 0.0207%, 0.0096%, and 0.0079% starch remaining, respectively. The negative control had no GA exposure, and had a higher (0.0247%) percentage of its starch remaining.~~

wc: ^starch

too much detail

The starch remaining per gram of seed was calculated to eliminate any error as a result of the varying weights of the sample. To determine the amount of starch digested per gram of seeds for each treatment, the percentage of starch digested in each sample was divided by its weight in grams (Figure 2).

move this ¶ ahead of previous one

fig format

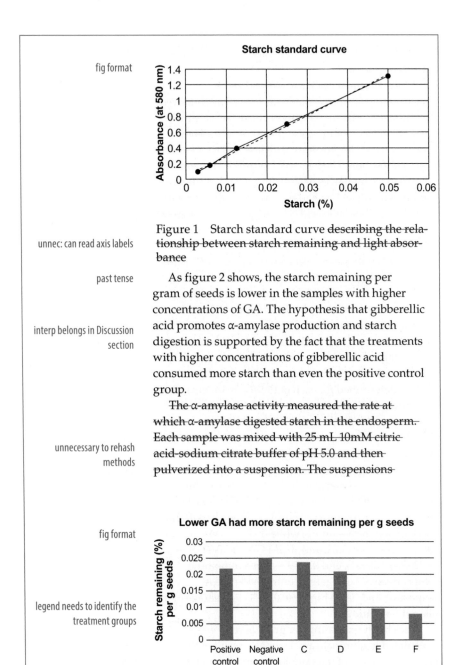

Starch standard curve

Figure 1   Starch standard curve ~~describing the relationship between starch remaining and light absorbance~~

unnec: can read axis labels

past tense

interp belongs in Discussion section

    As figure 2 shows, the starch remaining per gram of seeds is lower in the samples with higher concentrations of GA. The hypothesis that gibberellic acid promotes α-amylase production and starch digestion is supported by the fact that the treatments with higher concentrations of gibberellic acid consumed more starch than even the positive control group.

unnecessary to rehash methods

    ~~The α-amylase activity measured the rate at which α-amylase digested starch in the endosperm. Each sample was mixed with 25 mL 10mM citric acid-sodium citrate buffer of pH 5.0 and then pulverized into a suspension. The suspensions~~

fig format

Lower GA had more starch remaining per g seeds

legend needs to identify the treatment groups

to make title self-explanatory: ^GA ^barley

Figure 2   Effect of ^ treatments on the percentage of starch remaining per gram of ^ seeds

~~were centrifuged to derive supernatants containing the enzymes created in the reaction with citric acid-sodium citrate buffer. The supernatants of each sample were treated with 1 mL of 0.025% starch solution each and allowed to react for 120 seconds. The alpha amylase enzymes in the suspensions digested the starch solutions. The α-amylase activity was calculated by dividing the starch digested per gram of seeds by the reaction time (Figure 3).~~

meaning

The data in Figure 3 ~~show that the amount of starch digested was dependent on the amount of α-amylase present in the supernatant sample.~~ Treatments with higher concentrations of gibberellic acid ~~appeared to~~ have ^ the most α-amylase activity ^. The negative control had the least amount of activity at 3.16×10^{-6} starch digested/g/seeds/sec. This greatly contrasts with sample F, with 3.5 mL 10 μM GA and 1.43×10^{-4} starch digested/g/seeds/sec. The positive control also had a relatively low amount of α-amylase activity compared to the high concentration treatments. ~~The trend in Figure 3 shows that the α-amylase activity increased proportionately with the concentration of GA in each sample. The results shown in Figure 3 support the notion that GA content indirectly affects starch digestion.~~

meaning

past tense: ^had

^( )

super exp

rep

meaning

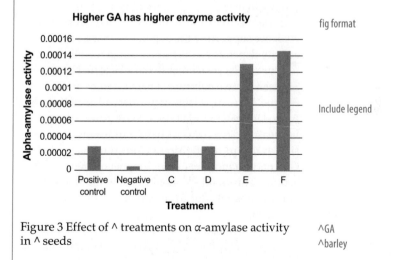

Higher GA has higher enzyme activity

fig format

Include legend

Figure 3 Effect of ^ treatments on α-amylase activity in ^ seeds

^GA

^barley

Discussion

wordy

The results shown in Figure 2 support the idea that amount of exogenous GA affects starch digestion in grass seeds through its induction of α-amylase production. As seen in Figure 2, higher concentrations of GA allowed seeds to digest higher percentages of starch. Samples E and F, with higher GA concentrations of 1 μM and 10μM, digested the

num format (use sci notat)

highest percentages of starch at .015495% and .0171%, respectively.

move this ¶ after next one. Disc structure specific to broad.

The experiment involved supplying endosperms with varying amounts of GA in order to simulate the embryo's natural production of α-amylase. The α-amylase enzyme is created in the aleurone layer of the seed and digests the available starches in the endosperm. The starches are broken down into glucose and nourish the seed in germination. Without gibberellic acid, the seed would not be able to process the available starches for nourishment.

The concentration of exogenous GA directly affected the α-amylase activity of each sample as well. Figure 3 shows that α-amylase activity was higher in samples with higher concentrations of exogenous GA. α-amylase activity increased according to the increase in GA. Sample C, with a 3.5 mL .01 μM GA concentration, yielded lower levels of α-amylase activity than even the positive control. Sample C's low α-amylase activity due to highly diluted GA provides evidence supporting the hypothesis that α-amylase activity depends on the presence of gibberellic acid.

Our results are similar to those reported by Skadsen for two cultivars of barley, Morex and Steptoe. He set up plates with embryo-less half seeds treated with 0, .01 μM and 1 μM GA. After 5 days he measured α-amylase activity. His results were similar to ours for the Morex cultivar in that higher GA concentrations resulted in higher α-amylase activity. This trend was not evident in the Steptoe cultivar, possibly due to an inhibitor in the endosperm

cit format

(Skadsen).

Production of α-amylase and starch degradation in the seed may have been affected by several sources of error. The separation of endosperm from embryo was performed without careful measurements; any embryo left attached to its endosperm may provide the seed with endogenous GA not accounted for in the calculations. In the negative control group, the presence of embryonic remains may have contributed to a slight amount of α-amylase activity (3.16×10^{-6} starch digested/g/seeds/sec) observed in Figure 3. Another factor contributing to error includes timing of each reaction with starch. The reaction began when 2 mL of .025% starch was added to each sample. Each reaction was intended to endure 120 seconds until stopped by the addition of 1 N HCl. If the 1 N HCl was not added at the correct time, the amount of actual starch digested would be inconsistent with the calculated value.

good error analysis

super exp

References

Koning, Ross E. 1994. Seeds and Seed Germination. Plant Physiology Information Website. http://plantphys.info/plants_human/seedgerm.shtml. (accessed 11-20-2016).

Paleg L.G. Physiological effects of gibberellic acid. II. On starch hydrolyzing enzymes of barley endosperm. Plant Physiology 1960; 35(6): 902-906.

cit missing

Skadsen RW. ^ Aleurones from a Barley with Low α-Amylase Activity Become Highly Responsive to Gibberellin When Detached from the Starchy Endosperm. Plant Physiol. (1993) 102: 195-203

N-Y ref format: year follows author's name ^1993

Write out journal title

Tanaka Y, Akazawa T (Plant Physiology) "Alpha-amylase isozymes in gibberellic acid-treated barley half-seeds" (1970) Apr; 46(4): 586^

ref format: year follows authors' names

^-591; inclusive pages

Check Your Understanding

1. Compare a section (abstract, introduction, materials and methods, results, or discussion) of the good with the not-so-good lab report in this chapter. Describe the features that characterize the good lab report. What are the main deficiencies of the corresponding section of the not-so-good lab report?

2. Prepare a checklist for the section you analyzed in (1). What information needs to be in this section? What information should not be in this section?

3. Apply your checklist to the corresponding section of a journal article recommended by your instructor. How well does the journal article meet the requirements of your checklist? Is there content that you think doesn't belong? Evaluate the figure format. Do you have any suggestions for improvement?

Go to **macmillanlearning.com/knisely6e**
and select **"Student Site"** to access
samples, template files, and tutorial videos

## Chapter 9

Poster Presentations

Objectives

9.1 Understand how the purpose of a poster differs from that of a scientific paper
9.2 Describe the characteristics of a well-designed scientific poster
9.3 Know how to create a poster in Microsoft PowerPoint
9.4 Understand how the traditional IMRD format is modified for posters
9.5 Summarize the poster for visitors in 3-5 minutes

Posters are a means of communicating preliminary research results both orally and visually. **Poster sessions** are often held in large exhibit halls at national meetings, where presenters stand next to their poster and discuss their work with interested attendees. The informal conversations provide great networking opportunities for young scientists.

9.1 Why Posters?

Scientists who attend poster sessions constitute a much larger audience than the one attracted to a journal article on a particular topic. Thus, your goal is to produce a poster that not only attracts experts in your subdiscipline, but also the much larger group of scientists with tangential research interests. The latter group provides a unique opportunity for you to learn about applications of your work to other research areas (and vice versa), spurs scientific creativity, and prompts you to apply an interdisciplinary approach to problem-solving.

Posters are *not* papers; they rely more on visuals than on text to present the message. It is not necessary to supply as many supporting details as you would for a paper, because you (the author) will be present to discuss details one-on-one with interested individuals. Too much material may even discourage individuals from reading your poster.

An appropriate poster presentation should fulfill two objectives. First, it must be esthetically pleasing to attract viewers in the first place. Second, it must communicate the objectives, methods, results, and conclusions clearly and concisely.

9.2 Poster Format

Posters can be created in Adobe InDesign, Microsoft Publisher, Adobe Illustrator, and Microsoft PowerPoint, among other programs, and printed on large-format paper. If you are presenting at a professional society meeting, check with the conference organizer regarding size and other details. For poster sessions in your class, ask your instructor about the requirements.

Layout

Make a rough sketch of the layout before you begin. Position the title and authors' names prominently at the top of the poster. Then arrange the body of the poster in 2, 3, or 4 columns, depending on whether the orientation of the poster is portrait or landscape. If you don't want to make your poster from scratch, check if your institution provides poster templates. Scientific poster templates are also available on the internet (see, for example, Alley [2020], Purrington [2020], and other sources listed in the Bibliography).

Appearance

The success of a poster presentation depends on its ability to attract people from across the room. Interesting graphics and a pleasant color scheme are good attention-getters, but avoid "cute" gimmicks. Present your poster in a serious and professional manner so people will take your conclusions seriously.

Organize the sections so that information flows from top left to bottom right. Align text on the left rather than centering it. The smooth left edge provides the reader with a strong visual guide through the material.

Avoid crowding. Large blocks of text turn off viewers; instead, use bullets to present your objectives and conclusions clearly and concisely. Use blank space to separate sections and to organize your poster for optimal flow from one section to the next.

Make sure your text is legible. Use black font on a white or light-colored background for best contrast.

Use appropriate graphics that communicate your data clearly. Use three-dimensional graphs *only* for three-dimensional data. Make sure that the numbers and labels on figures are large and legible and free of typos.

Font (type size and appearance)

Keep in mind that most visitors will be reading your poster from 3 to 6 feet away. Sans serif fonts like **Arial** are good for titles and labels, but serif fonts like **Times** and **Palatino** are much easier to read in extended blocks of text. The **serifs** (small strokes that embellish the character at the top and bottom) create a strong horizontal emphasis, which helps the eye scan lines of text more easily.

Format the title of the poster in sentence case or title case in 72 point or larger **bold**. In **sentence case**, only the first letter of the first word is capitalized. In **title case**, the first letter of the most important words is capitalized. These styles are easier to read and take up less space than all caps. Keep the case of case-sensitive words, such as pH, cDNA, or mRNA, as is. Limit title lines to 65 characters or less.

Times 72

Arial 72

Authors' names and affiliations should be at least 48 point **bold**, serif font, title case:

Times 48 Pt

The section headings can be 32-48 point **bold**, serif font, sentence case or title case:

Times 28 Pt

The text itself should be no smaller than 24 point, serif font, sentence case, and *not* in bold:

Times 24 pt

In terms of readability, it's better for the font to be larger rather than smaller.

Nuts and bolts

Ask the conference organizer (or your instructor) about how posters will be displayed at the poster session. Some possibilities include a pinch clamp on a pole, an easel, a table for self-standing posters, and cork bulletin boards to which posters are attached with pushpins.

9.3 Making a Poster in Microsoft PowerPoint

Many scientists use PowerPoint to prepare posters for professional society meetings because this software is readily available; the font, color scheme, and layout can be customized by the authors; the content can be revised easily; the electronic file format facilitates collaboration among colleagues in different locations; and the final product looks professionally done. Because you are at the mercy of your computer when making posters in PowerPoint, however, save your file early and often!

The entire poster is made on one PowerPoint slide, whose size is set in the **Slide Size** dialog box. Text is written in text boxes, which are inserted along with images and graphs on the slide as desired. The following sections describe some of the basic tasks you will carry out when making your poster in PowerPoint.

Design

1. Open a blank PowerPoint presentation.

2. Click **Design | Slide Size**. Windows users click **Custom Slide Size**; Mac users click **Page Setup**. Then enter the poster dimensions (Figure 9.1). Stay within the dimensions specified by the conference organizer, and check with whomever is printing your poster about any size restrictions. Common sizes include 30×40″, 36×48″, and 56×41″. Portrait or landscape orientation will be selected automatically based on how you enter the numbers. An advantage of landscape is that visitors don't have to look up or down as far to view the whole poster.

3. Use black text on a white or light-colored background for best contrast.

Adding text, images, and graphs

- To add text to your poster, click **Insert | Text Box** for each section or block of text. Aim for a consistent look by using the same family of fonts. Adjust the font size for the title, authors' names and affiliations, section headings, and text by making the appropriate selections on **Home | Font**. To change the properties of the text box

(A)

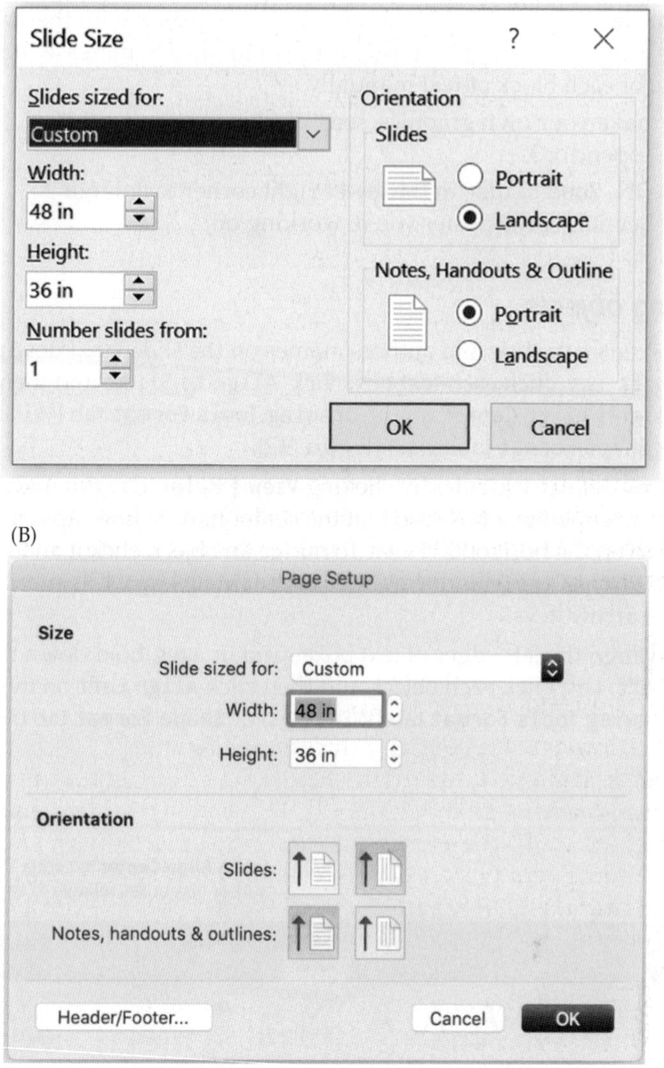

(B)

Figure 9.1 Poster dimensions are specified in the **Slide Size** dialog box on the **Design** tab. (A) PowerPoint 2019. (B) PowerPoint for Mac 2019.

itself, Windows users click the **Drawing Tools Format** tab on the Ribbon. Mac users click **Shape Format**.

- To insert pictures, click **Insert | Pictures**.

- To insert a graph, copy it from Excel and paste it into your Power-Point slide as a picture. When you drag on one of the corners, the font size of the axis labels automatically increases with the size of the graph. With regular paste, you would have to increase the font size of each block of text manually.
- To make your own graphics, see "Working with shapes" in Appendix 3.
- Use the **Zoom slider** in the lower right corner to enlarge the section of the poster you're working on.

Aligning objects

1. To center the title and authors' names on the slide, hold down the **Shift** key, click each text box, click **Align to Slide** and then **Align | Align Center** on the **Drawing Tools Format** tab (Windows) or **Shape Format** tab (Mac) (Figure 9.2).

2. Now display the ruler by clicking **View | Ruler**. Drag or nudge the grouped text boxes so that the center handle lines up with the zero on the horizontal ruler. To nudge an object, click it and hold down the **Ctrl** (Windows), or **Command** (Mac) key while pressing the arrow keys.

3. To align the left edges of text boxes and images, hold down the **Shift** key, click each object, and then click **Align Left** on the **Drawing Tools Format** tab (Windows) or **Shape Format** tab (Mac).

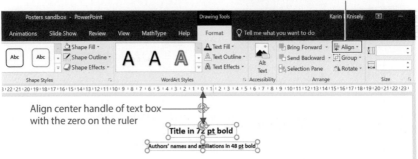

Select **Align Center** to center the title and author text boxes relative to each other

Align center handle of text box with the zero on the ruler

Title in 72 pt bold

Authors' names and affiliations in 48 pt bold

Figure 9.2 Center the title and authors' names and affiliations by first clicking **Align Center** and then aligning the center handle of the text box with the zero on the horizontal ruler. This screenshot is from PowerPoint 2019. The menus for PowerPoint for Mac 2019 are similar.

Proofread your work

Print a rough draft of your poster on an 8.5×11″ sheet of paper. To do so, click **File | Print**. Windows users, under **Settings**, click the down arrow next to **Full Page Slides** and check that **Scale to Fit Paper** is selected (this is the default for both Windows and Mac). Check for typos and grammatical errors and evaluate the layout and overall appearance of your poster. If the print is not large enough to read on the draft, it will probably be hard to read from a distance on the full-sized poster as well. Adjust the font size under **Home | Font**.

Final printing

Check with your local print shop concerning the electronic format of the poster file. Portable Document Format files (PDFs) are the printing industry standard. PDFs tend to be smaller than the source file, and all of the fonts, images, and formatting are retained when the document is printed.

9.4 Poster Content

Posters presented at professional society meetings should be organized so that readers can stand 10 feet away from the poster and get the take-home message in 30 seconds or less. Because of the large number of sessions (lectures) and an even larger number of posters, conference participants often experience "sensory overload." Thus, if you want your poster to stand out, make the section headings descriptive, the content brief and to the point, and the conclusions assertive and clear.

Posters for a student audience in the context of an in-house presentation should follow the same principle of brevity, but may retain the sections traditionally found in scientific papers. These include:

- Title banner
- Abstract (optional—if present, it is a summary of the work presented in the poster)
- Introduction
- Materials and Methods
- Results
- Discussion or Conclusions
- References (less comprehensive than in a paper)
- Acknowledgments

Title banner

Use a short, yet descriptive, title. This is the first and most important section for attracting viewers, so try to incorporate the takeaway message in the title. For example, **"One Big, Smelly Family: Decoding the Olfactory Receptors in the Indian Jumping Ant"** (see poster by Naughton et al. 2020 at **macmillanlearning.com/knisely6e**) conveys humor and piques your curiosity more than its journal article counterpart "**Evolution, developmental expression and function of odorant receptors in insects**" (Yan et al. 2020).

The title banner should be at the top of the poster and in 72 point (or larger) bold font, sentence case or title case, 65 characters or less on a line. Underneath the title, include the authors' names and the institutional affiliation(s). Use at least 48 point bold, title case for the authors' names.

Introduction

Instead of using the conventional "Introduction" heading found in papers, use a short, attention-grabbing question such as "How do ants communicate?" Then provide the minimum amount of background information on the topic and state your hypothesis. A bulleted list may be a good way to present some of this information.

Materials and methods

Summarize the methods in enough detail so that someone with basic training could understand your experiment. If possible, use pictures to show the experimental setup and flow diagrams or a bulleted list for the procedure.

Results

The Results section of a poster consists mostly of visuals (pictures and graphs). Announce each important result with a heading *above* the figure. It is not necessary to write a detailed caption below the figure, as you would in a scientific paper. Use a bulleted list to point out general patterns, trends, and differences in the results. Poster viewers do not have the time to read the results leisurely, as they would with a paper.

Figures may contain some statistical information including means, standard error, and minimum and maximum values, where appropriate. Make the data points prominent and use a simple vertical line without crossbars for the error bars. Make the font size on the axis numbers and axis titles at least 24 pt. When there are multiple data sets on a graph, instead of using a legend, place a descriptive label next to each line. This approach allows visitors to identify the conditions more quickly. Avoid using tables with large amounts of data.

Images are a great addition to a poster, but only if they are sharp. Picture quality is determined by the size (number of pixels) of the digital image and the printer resolution. Various sources recommend a printer resolution of at least 240 pixels per inch (ppi) to print quality photos. That means for a 5"×7" print, the image would have to be at least 1200×1680 pixels to attain the desired resolution on paper. To check the size of an image, right-click it and select **Properties**. Be careful when resizing pictures. Enlarging an image that does not have enough pixels will result in a blurry picture.

Edit the text ruthlessly to remove nonessential information about the visuals. A sentence like "Different odorant receptors were sensitive to different volatile chemicals" is nothing but deadwood. On the other hand, "The HsOr196 receptor was more sensitive than the HsOr70 receptor to odorants with a lower carbon count" informs the reader of the result.

Remember to leave blank space on a poster. Space can be used to separate sections and gives the eyes a rest.

Discussion or Conclusions

Most visitors will be especially interested in the Conclusions section, so make sure this section is featured prominently on your poster. This is where you discuss your results. Use bullets to state the take-home message and provide the experimental evidence. State whether or not your results support your hypothesis. Were there any unexpected or surprising results? What questions remain unanswered? If you were to repeat the study, what would you do differently?

Literature citations

In posters as in scientific papers, it is imperative to acknowledge your sources, which are cited most often in the Introduction and Discussion/ Conclusions sections. Because posters are informal presentations and the author is present to provide supporting details, the References section of a poster can be less comprehensive than that in a paper. Some ways to save space while still acknowledging your sources are to use the citation-sequence system and to abbreviate the end references, for example, by eliminating the article title or by using DOIs and URLs for online publications. However, for in-class poster presentations, your instructor may ask you to hand in a correctly formatted reference list or bibliography. For in-text and end reference format, see Chapter 4.

Acknowledgments

In this section, the author acknowledges organization(s) that provided funding and thanks technicians, colleagues, and others who have made significant contributions to the work.

9.5 Presenting Your Poster

Prepare a 3–5 minute talk explaining your poster. This "elevator pitch" should be an engaging summary of why and how you did the work, your most important findings, and your conclusions. Even if you don't actually give your pitch at the poster session, verbalizing the content of your poster is good preparation for the one-on-one conversations you will have with conference participants. Bring along a stack of business cards or handouts with your contact information to give to interested visitors for possible future collaboration.

9.6 Evaluation Form for Poster Presentations

Good posters are the product of creativity, hard work, and feedback at various stages of the poster preparation process. When you present a poster to your class or at a professional society meeting, participants may be asked to evaluate your poster using a form such as the "Evaluation Form for Poster Presentations" available at **macmillanlearning.com/knisely6e**.

When you are in the position of evaluator, make the kinds of comments you would find helpful if you were the presenter. As you know, feedback is most likely to be appreciated when it is constructive, specific, and done in an atmosphere of mutual respect.

9.7 Sample Posters

Examples of posters along with reviewer comments are posted on **macmillanlearning.com/knisely6e**. What do you like (or not like) about each poster? Use the evaluation form to determine how well the authors have met the requirements of good poster design.

> Go to **macmillanlearning.com/knisely6e**
> and select **"Student Site"** to access
> samples, template files, and tutorial videos

## Chapter 10

Oral Presentations

Objectives

10.1 Learn how the structure of oral presentations differs from that of scientific papers

10.2 Understand how your audience, purpose, and speaking environment affect your presentation

10.3 Describe features of slides that are easy for an audience to process

10.4 Describe features of slides to avoid, because they are hard for an audience to process

10.5 Understand the importance of preparation and rehearsal for a successful presentation

10.6 Recognize and emulate the qualities of excellent speakers

10.7 Understand how slide decks can be repurposed as reference materials

Scientific findings are communicated through journal articles, poster presentations at meetings, and oral presentations. Oral presentations are different from journal articles and posters because the speaker's delivery plays a critical role in the success of the communication.

As a student, you have been on the receiving end of oral presentations for a number of years and probably have a pretty good idea of what makes a talk memorable. For example, presentations in which the speaker appears to be unprepared, uncomfortable speaking in public, and unfamiliar with the level of his or her audience are quickly forgotten regardless of the content. On the other hand, successful presentations tend to have the following characteristics:

- Speakers who show enthusiasm for their topic
- Speakers who establish a good rapport with the audience
- Information that is well organized, leaving listeners satisfied that they understand more than they did before
- Visuals that are simple and legible, and help listeners focus on the important points without drawing attention away from the speaker

10.1 Structure

Oral presentations are not unlike scientific papers in their structure, but they are much more selective in their content. As in a scientific paper, the **introduction** captures the audience's attention, provides background information on the subject, and identifies the objectives of the work (Table 10.1). The **body** is a condensed version of the Materials and Methods and Results sections, and contains only enough detail to support the speaker's conclusions. If the focus of the talk is on the results, then the speaker spends less time on the methods and more time on visuals that highlight the findings. The visuals should be simpler than those prepared for a journal article, because the audience may only have a minute or two to digest the material in the visual (in contrast to a journal article, where the reader can re-read the paper as often as desired). Finally, the **closing** is comparable to the Discussion section, because here the speaker summarizes the objectives and results, states conclusions, and emphasizes the takeaway message for the audience. The closing may include an acknowledgments slide on which the speaker lists sources of funding and recognizes research advisers, collaborators, and others who helped with the work.

There is no distinct abstract or References section in an oral presentation. If the presentation is part of a meeting, an abstract may be provided in the program. References may be given in an abbreviated form at the bottom of a slide or below visuals. If the **slide deck** (the slides with any notes) will be shared, full references can be included in the notes.

10.2 Plan Ahead

Before you start writing the content of your presentation, make sure you know the following:

- **Your audience.** What do they know? What do they want to know? If possible, ask some carefully chosen questions to assess their experience and motivation.

TABLE 10.1   Comparison of the structure of an oral presentation and a journal article

Oral Presentation	Journal Article
	Abstract
Introduction	Introduction
Body	Materials and Methods, Results
Closing	Discussion
	References
	Acknowledgments

- **Why you are giving the talk**, not just why you did the experiment. Talks may be **instructional**—the speaker wants listeners to leave with new knowledge and skills. Talks may be **informational**—the speaker presents research to fellow scientists at a seminar or journal club meeting. Or talks may be **persuasive**—the speaker may try to get listeners to provide funding or to adopt a certain point of view on a controversial topic.
- **Your speaking environment.** How much time do you have? How large is the audience? What presentation equipment is available (chalkboard, overhead projector, computer with projection equipment, internet connection)?
- **How the slide deck will be used** beyond the presentation. If you intend to share the deck, include explanatory text that would enable your remote audience to understand the slides without you, the speaker, being present.

10.3 Prepare the First Draft

An oral presentation will actually take you longer to prepare than a paper. D'Arcy (1998) breaks down the steps as follows:

- Procrastination   25%
- Research   30%
- Writing and creating visuals   40%
- Rehearsal   5%

To overcome procrastination, follow the same strategy as for writing a laboratory report: write the body first (Materials and Methods and Results sections), then the introduction, and finally the closing.

- Write the main points of each section in outline form.
- Use science news articles, review articles, and journal articles to provide background in the introduction and supporting references in the closing. Use sources that are appropriate and engaging for your audience.
- Determine how much detail to present on each topic. Consider the level of your audience and how much time you have for the presentation.
- Transfer each topic on your outline to its own slide in your presentation software program. There are several programs on the market, but Microsoft PowerPoint is the most popular. PowerPoint has been widely criticized for its bullet-laden templates, but it is really up to you, the presenter, to choose a suitable layout and design for your slides.
- Do a preliminary, timed run through your presentation. Eliminate nonessential information to stay within the time limit.

10.4 Make the Slides Audience-Friendly

The first draft of your presentation is all about you, the speaker. The topic-bullet approach is a convenient way for you to organize your talk, but text-heavy slides are a sure way to turn off your audience. Audience members simply cannot read line after line of incomplete thoughts and simultaneously listen to what you are saying. Therefore, it is imperative that you make your slides audience-friendly. Studies have shown that a combination of *hearing* the speaker, *reading* a simple, complete statement, and *seeing* a visual increases audience comprehension and retention (Alley 2016; Nathans-Kelly and Nicometo 2014; McConnell 2011).

Here are some suggestions for making your slides audience-friendly. Of course, if you were given specific instructions for your presentation, then those take precedence over these suggestions.

Focus on one idea at a time

Evaluate each slide critically. Ask yourself, "What is the point of this slide?" If you find yourself listing several key points, consider splitting up the information over several slides. **One idea per slide** is easier for listeners to comprehend, because they can focus all their attention on just one topic, instead of having to figure out how multiple topics are related.

Write complete thoughts

Replace ambiguous slide titles like "Introduction" with complete thoughts that eliminate guesswork. Give audience members the takeaway message; don't make them waste valuable time wondering what they are supposed to remember (Figure 10.1). The **complete-sentence style titles** should be written in active voice and sentence case and have consistent punctuation and alignment. Titles should be no longer than two lines when written in 32–40 point font.

Use more visuals and fewer words

As much as possible, **replace text with a visual** that supports the title sentence (Figure 10.2). It takes less time and effort to comprehend a picture than to read and process the words that the picture illustrates. Be careful not to overwhelm the audience with details. Present the minimum amount of information that listeners need to understand the topic.

Keep graphs simple

Make the numbers and the lettering on the axis labels large and legible. Choose symbols that are easy to distinguish and use them consistently in

(A) Script for the speaker

(B) Audience-centered slide

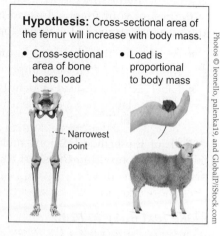

Introduction

- The femur is an unpaired bone that has to bear the entire weight of an animal during running.

- Load is proportional to mass.

- Cross-sectional area has to bear the load.

- We wanted to see how the cross-sectional area changes with the mass of the animal.

Hypothesis: Cross-sectional area of the femur will increase with body mass.

- Cross-sectional area of bone bears load

- Load is proportional to body mass

Narrowest point

Photos © leonello, palenka19, and GlobalP/iStock.com

Figure 10.1 Speaker-friendly (A) and audience-friendly (B) ways to construct a slide. The large amount of text on Slide A will have the audience reading instead of listening to the speaker. The simple, complete statement on Slide B tells the audience what is important. Well-chosen pictures, displayed sequentially for emphasis, support the statement.

(A) Script for the speaker

(B) Audience-centered slide

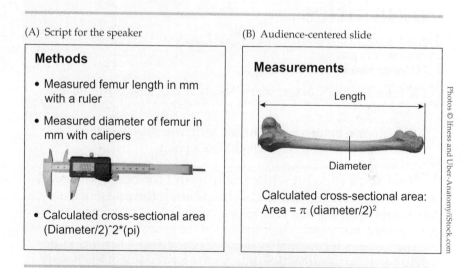

Methods

- Measured femur length in mm with a ruler

- Measured diameter of femur in mm with calipers

- Calculated cross-sectional area (Diameter/2)^2*(pi)

Measurements

Length

Diameter

Calculated cross-sectional area:
Area = π (diameter/2)^2

Photos © Ifness and Uber-Anatomy/iStock.com

Figure 10.2 Speaker-friendly (A) and audience-friendly (B) ways to construct a slide. The audience can look at a simple, labeled picture and understand what was measured much faster than reading a description of the procedure. Irrelevant details, such as the instrument used to make the measurements, are omitted from the slide.

all of your figures. For example, if you use a circle to represent "Wild type in glycerol" in the first graph, use the same symbol for the same condition in the rest of your graphs. Instead of following the conventions used in journal articles, modify the figure format as follows (Figure 10.3):

- Position the title above the figure and don't number it. The title should reflect your most important finding.
- Label each line instead of identifying the data symbols with a legend. The labels may be made using text boxes in Excel, Word, or PowerPoint.

Both of these modifications make it easier for the listener to digest the information within the short time the graph is displayed during the presentation.

Make text easy to read

Use fonts and a color scheme to make text easy to read from anywhere in the room.

Use at least 24 pt font for bullet points, labels, and axis titles

Use sans serif font for slides for easier reading.

Use serif font for handouts.

Use color sparingly in text slides: Nearly 10% of the male population is color blind.

A bright color, **boldface**, or a larger font, when used with discretion, are good ways to emphasize a point.

Use black type on white background for best contrast. The light background also helps keep the audience awake and allows you to see your listeners' faces.

In a large room, light-colored type on a dark background makes characters look bigger.

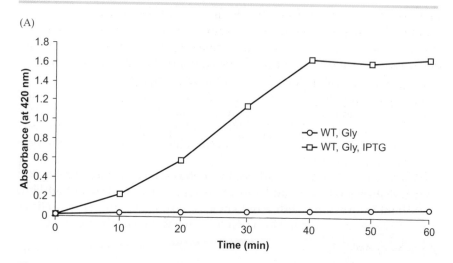

Figure 1 Beta-galactosidase production in wild type *E. coli* grown in glycerol media. Inducer (IPTG) was added to the culture at the start of the assay.

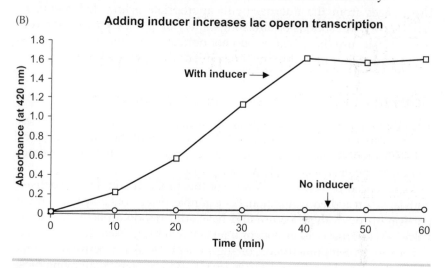

Figure 10.3   Example of figure formatting for (A) a journal article and (B) an oral presentation.

Provide ample white space

Use white space liberally, especially around the edges of the slide. If the slides are formatted for wide screen (16:9) and the projection equipment is set for standard format (4:3), content near the edges may be cut off.

Use animation feature to build slide content

Sometimes it may be necessary or desirable to present a complicated concept, process, or result on a single slide. However, rather than displaying all of the components at once, the speaker "builds" content by revealing only one element at a time. This technique is often applied to overview and conclusion slides, on which bulleted items appear one after the other, allowing the audience to focus on one point at a time while the speaker is talking about it. This effect is produced with PowerPoint's animation feature. For instructions on how to apply animations to your PowerPoint presentations, see Section A3.3.

Shapes can also be animated to add emphasis. For example, the speaker may call attention to a certain point on a graph with an arrow or highlight important information with a red circle or box. Using animations eliminates any ambiguity caused by a pointer being waved in the general direction of the visual.

Appeal to your listeners' multiple senses

Use images from the internet (with appropriate acknowledgment), recordings of heart sounds or bird songs, and videos if they help make your point. Make sure these "extras" do not detract from your takeaway message and that they fit within the time limit of your presentation.

Deal proactively with lapses in audience attention

The attention level of an audience is highest at the beginning of a presentation, decreases in the middle, and then recovers at the end. In a typical scientific presentation, that means the audience will be the most distracted during the most technical part of the talk.

Experienced presenters are aware of this problem and deal proactively with lapses in audience attention. For example, professors may take a break partway through their lecture and have students do an activity that involves small-group discussion, or a reflective writing exercise, or a clicker quiz to assess comprehension. Forcing students to take an active role in the class has the dual benefit of clearing up misconceptions and allowing those who zoned out to catch up. In research talks, you may have noticed that after speakers present an especially complex topic, they pause and review the most important points. These approaches for recapturing audience attention are analogous to a diver (the speaker) swimming to deep waters and then returning to the surface after a while, as illustrated in Figure 10.4. The idea is that the speaker dives into increasingly difficult material, but periodically brings the audience up for air. During this recovery phase, the audience gets another chance to hear the key points, and the respite helps prevent those whose attention has lapsed from becoming hopelessly lost.

Figure 10.4 "Data dives" approach for keeping the audience on track during the most technical portion of an oral presentation. The speaker-diver takes audience members into depth on a topic, but then periodically brings them up for air by recapping the most important ideas. (Adapted from S. McConnell. 2011. *Designing effective scientific presentations.* [uploaded 13 Jan 2011; access 2016 Dec 18]. https://www.youtube.com/watch?v=Hp7Id3Yb9XQ.)

10.5 Rehearsal

After you have prepared your presentation, you must practice your delivery. Give yourself plenty of time so that you can run through your presentation several times and, if possible, do a practice presentation in the same room where you will hold the actual presentation. Here are some other tips:

- Go to a place where you can be alone and undisturbed. Read your presentation aloud, paying attention to the organization and especially the connecting sentences. Does one topic flow into the next, or are there awkward transitions? Revise both the PowerPoint and your speaker notes as needed. Apply the revision strategies described in Chapter 7 to your oral presentation.

- Time yourself. Make sure your presentation does not exceed the time limit.

- Practice, practice, practice. With each round of practice, try to rely less and less on your notes.

- After you are satisfied with the organization, flow, and length, ask a friend to listen to your presentation. Ask him or her to evaluate your poise, posture, voice (clarity, volume, and rate), gestures and mannerisms, and interaction with the audience (eye contact, ability to recognize if the audience is following your talk).

It is natural to be nervous when you begin speaking to an audience, even an audience of your classmates. Adequate knowledge of the subject, good preparation, and sufficient rehearsal can all help to reduce your nervousness and enhance your self-assurance.

10.6 Delivery

The importance of the delivery cannot be overstated: listeners pay more attention to body language (50%) and voice (30%) than to the content (20%) (Fegert et al. 2002). Remember that you must establish a good rapport with the audience in order for your oral presentation to be successful. The following guidelines will help.

Presentation style

- Dress appropriately for the occasion.
- Use good posture. Good posture is equated with self-assurance, while slouching implies a lack of confidence.
- Be positive and enthusiastic about your subject.
- Try to maintain eye contact with some members of the audience. In the United States, eye contact is perceived as more personal, as though the speaker is having a conversation with individual listeners.
- Avoid distracting gestures and mannerisms such as pacing, fidgeting with the pointer, jingling the change in your pocket, and adjusting your hair and clothes.
- Speak clearly, at a rate that is neither too fast nor too slow, and make sure your voice carries to the back of the room. Avoid "um," "ah," "like," and other nervous sounds.
- Do not stand behind the computer monitor and read your presentation off the computer screen. Similarly, do not turn your back on the audience and read your presentation off the projection screen. Both practices distance you from your audience and make you seem unprepared.
- If you use notecards, number them so that if you drop them you can reassemble them quickly.
- Do not display a visual until you are ready to discuss it.
- When pointing to something on the projection screen, stand close enough to point with your outstretched arm or pointer (stick) or use a laser pointer.
- Tell the audience exactly what to look for. Interpret statistics and numbers so that they are meaningful to your listeners. If you are describing Figure 10.3, for example, "Beta-galactosidase production increased dramatically in only 10 minutes after adding the inducer" is likely to make a bigger impression on the audience than "From 0 to 10 minutes, the absorbance increased from 0.01 to 0.24."

- Point to the specific part of the visual that illustrates what you are describing. Avoid waving the pointer in the general direction of the visual.
- Do not block the projected image with your body.

Interacting with the audience

A unique benefit of communicating information orally is that you can adjust your presentation based on the feedback you receive from your listeners. Pay attention to the reaction of individuals in the audience. Does their posture or facial expression suggest interest, boredom, or confusion? If you perceive that the audience is not following you, ask a question or two, and adjust your delivery according to their response. Your willingness to customize your presentation to your audience will enhance your reputation and lead to greater listener satisfaction.

To keep listeners focused:

- Establish common ground immediately. Ask a question or describe a general phenomenon with which your audience is familiar.
- Incorporate humor if you can do so without being culturally insensitive or offensive.
- Make sure your introduction is well organized and proceeds logically from the general to the specific without any sidetracks.
- Alert your audience when you plan to change topics. Practice these transitions when you rehearse.
- Summarize the main points from time to time. Tell your listeners what you want them to remember.
- Use examples that are relevant to your audience. Keep in mind that older members of the audience may not be familiar with the latest viral videos on the internet. Conversely, younger members of the audience may not know what TV shows were popular in the 1980s.
- End your presentation with a definite statement such as "In conclusion..." Don't fade out weakly with phrases such as "Well, that's about it."

Group presentations

If you are presenting with another person, your individual contributions should complement one another, not compete with each other. This requires good coordination and practice beforehand.

Fielding listener questions

Allow time for questions and discussion. Take questions as a compliment—your listeners were paying attention! Listen carefully to each question, repeat it so that everyone in the room can hear, and then give a brief, thoughtful response. If you don't understand the question, ask the listener to rephrase it. If you don't know the answer, say so. You might offer to find out the answer and follow up with the person later.

Signal the end of the question session with "We have time for one more question" or something similar.

10.7 Slide Decks as References

A **slide deck** is the group of slides that makes up a presentation. Besides being used as visual aids during the presentation, slide decks have the potential to be valuable references in the absence of the speaker. Your instructors probably post their lecture slide decks on a learning management system such as Blackboard or Moodle for you to review. Research collaborators share slide decks to avoid having to recreate images and reexplain basic concepts. In the corporate world, slide decks enable those who are unable to attend a meeting to stay informed, have access to supporting documentation, and edit the slides for other presentations.

When the slide decks are conceived primarily as references instead of visual aids for the audience, invariably too much text ends up on the slides. A solution to this problem is to use the notes pane for explanatory text, and design the slide so that the audience can focus on the takeaway message. To open the notes pane in PowerPoint 2019 and PowerPoint for Mac 2019, click **Notes** at the bottom of the window. Type your talking points as bullets or running text; there is no limit to the length of the notes. Print out the notes pages or save them as a PDF:

- Windows: `File | Print | Settings | Print Layout | Notes Pages`
- Mac: `File | Print | Layout : Notes`

In this format, each page consists of a thumbnail of the slide along with the explanatory notes (Figure 10.5).

Notes can also be used as prompts for the speaker during the presentation. In **Presenter View**, the speaker can see the notes on the computer screen, but the audience sees only the current slide on the projection screen (Figure 10.6). Other features of **Presenter View** are described in Section A3.4.

Slide decks can be shared in a variety of ways, as shown in Table 10.2.

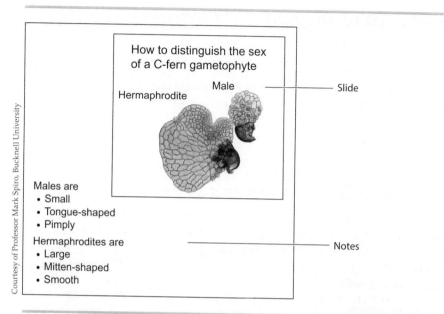

Figure 10.5 Slide decks saved as notes pages can be used as speaker notes during the presentation and as references before and after the presentation.

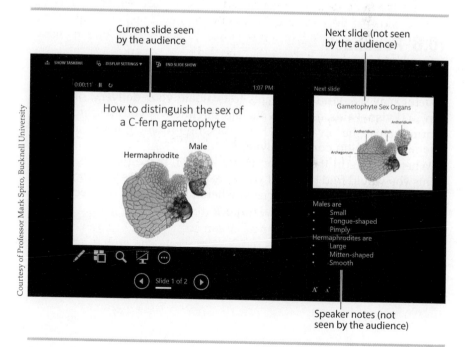

Figure 10.6 Presenter View in PowerPoint allows the speaker to view notes on the computer screen, eliminating the need to create text-heavy slides.

TABLE 10.2 Sharing slide decks depending on their purpose	
Purpose	Command
Handouts for notetaking. Print thumbnails of the slides (without notes) so that listeners can take notes during the presentation.	`File	Print	Settings	Handouts	3 Slides` Select printer for hard copies or print to PDF for sharing electronically.
Documentation. Print a thumbnail of the slide along with the notes.	`File	Print	Settings	Print Layout	Notes Pages` Select printer for hard copies or print to PDF for sharing electronically.
Reuse slides and play animations	`File	Save as` PowerPoint Presentation (.ppt or .pptx)
Narrate slide show and save as a video	See Section A3.4
Speaker prompts. View notes pane in Slide Show mode.	`Alt-F5` on a PC. `Option+Return` on a Mac.

10.8 Feedback

It takes not only practice, but also coaching, to become an effective speaker. When you make an oral presentation to your class, your instructor may ask your classmates to evaluate your delivery, organization, and visual aids, as well as the thoroughness of your research using a form such as the "Evaluation Form for Oral Presentations" available at **macmillanlearning.com/knisely6e**.

When you are in the position of the listener, make the kinds of comments you would find helpful if you were the speaker. As you know, feedback is most likely to be appreciated when it is constructive, specific, and done in an atmosphere of mutual respect.

Check Your Understanding

1. Watch videos about presenting scientific talks, which are recommended by your instructor or listed under "Oral Presentations" in the Bibliography (see Alley 2016 and McConnell 2011). Take notes on what characterizes an effective oral presentation for other scientists.

2. Watch one of the TED "Most popular talks of all time" (https://www.ted.com/playlists/171/the_most_popular_talks_of_all). Keep in mind that TED audiences tend to be educated, but are not necessarily experts on the topic.

   a. Evaluate the presentation style of the speaker. How does the speaker engage with the audience? How does the speaker keep the audience engaged? What techniques does the speaker use to emphasize important points?

   b. Evaluate the visuals. How do the text and images support the speaker's spoken words? Do they focus on one topic at a time?

> Go to **macmillanlearning.com/knisely6e**
> and select **"Student Site"** to access
> samples, template files, and tutorial videos

Word Processing in Microsoft Word

A1.1 Introduction

Microsoft Office 365 supports both the online and desktop versions of Microsoft Word 2019 for Windows and Word for Mac 2019. The online app is handy for accessing files from multiple devices and for collaborative writing. However, it does not have the full functionality of the desktop version of MS Word.

This appendix will address commands in the desktop version of MS Word, which are frequently used when writing scientific papers. See also the video tutorials for Windows and Mac at **macmillanlearning.com/knisely6e**. For all other word processing questions, please visit the Microsoft Office support website.

In this book, the nomenclature for the command sequence is as follows: **Ribbon tab | Group** (Windows only) **| Command button | Additional Commands** (if available). For example, in Windows, **Home | Font | Superscript** means "Select the **Home** tab on the Ribbon, and in the **Font** group, click the **Superscript** button" (Figure A1.1A). In the Mac OS, the corresponding command sequence is **Home | Superscript** (Figure A1.1B).

Figure A1.1 Command area in (A) Word 2019 and (B) Word for Mac 2019.

A1.2 Good Housekeeping

Organizing your files in folders

You can expect to type at least one major paper and perhaps several minor writing assignments in every college course each semester. This amounts to a fair number of documents on your computer. One way to organize your files is to make individual course folders that contain subfolders for lecture notes, homework, and lab assignments, for example. To create a new folder in Word when you are saving a file:

Mac

1. Click **File | Save As**. The **Save As** dialog box appears.

2. Click **Online Locations | +** to access **OneDrive** to save your file to Microsoft's cloud (you must have set up an account to use this option).

3. If you want to save the file to your computer or iCloud Drive, do *not* click **Online Locations**. Instead, click the down arrow next to the **Save As** box, and navigate to a preexisting folder or create a new folder.

4. To see the files you stored on iCloud Drive, click **Finder | iCloud Drive**.

5. To create a new folder, click the **New Folder** button. The **New Folder** dialog box appears.

6. Give the folder a short, descriptive name. Click **Create**.

To create a new folder in **Finder**:

1. Click **Finder**. Navigate to the desired location.

2. Click ✿ ▼ | **New Folder** on the menu bar. Type a name for the new folder and press **Return**.

Windows

1. Click **File | Save As**. The **Save As** dialog box appears.

2. Click **OneDrive** to save to Microsoft's cloud (you must have set up an account to use this option). If you want to save the file to your computer, navigate to a preexisting folder or create a new folder.

3. To create a new folder, click the **New Folder** button. The **New Folder** dialog box appears.

4. Give the folder a short, descriptive name. Click **OK**.

To create a new folder in **Windows Explorer**:

1. Click **Start | File Explorer | Documents**. The Documents library dialog box appears.

2. Click the **New folder** button in the middle of the **Home** tab menu. Type a name for the new folder and press **Enter**.

Accessing files and folders quickly

Mac Word makes recently used files readily accessible by displaying them when you click **File | Open Recent**. Frequently used files can also be pinned to the **Recent Documents** list:

1. Click **File | Open Recent | More**.
2. When you hold the cursor over a file, a thumbtack will appear on the right. Click the thumbtack to pin the file to the top of the list.
3. When you are finished working on the file, click **File | Open Recent | More**, and click the pinned file to remove it from the list.

Windows Word makes recently used files readily accessible by displaying them chronologically when you click **File | Open**. Frequently used files can also be pinned to the top of the list:

1. Click **File | Open**.
2. Right-click a file and select **Pin to list** from the menu. Or hover over the document and click the blue pin on the right.
3. The file will be pinned to the top of the list.
4. When you are finished working on the file, simply unpin the files.

To pin files or folders to **Quick access** in Windows Explorer:

1. Open a Windows Explorer dialog box by clicking **Start | File Explorer | Documents**.
2. Browse your documents library until you find the frequently used file (or folder).
3. Hold down the left mouse button and drag the file (or folder) to Quick access in the navigation pane on the left.

Naming your files

File names can include letters, numbers, underlines, and spaces as well as certain punctuation marks such as periods, commas, and hyphens. The following characters are not allowed: \, /, :, *, ?, ", <, >, and |.

In general, file names should be short and descriptive so that you can easily find the file on your computer. If you intend to share your file, however, consider the person you are sharing it with. When you send your lab report draft to your instructor for feedback, put your name and the topic in the file name (for example, Miller_lacoperon). Although "biolab1" may seem unambiguous to you, it is not exactly informative for your biology professor!

If you have forgotten the exact file name and location, you can search for it. On a Mac, use **Finder**. In Windows, click **Start | File Explorer | This PC | Windows (C:)** (or **Documents** if you are able to narrow down the location).

Saving your documents

Nothing is more frustrating than to spend hours typing a paper, only to lose the file when your computer crashes. Fortunately, Word automatically saves your file at certain intervals. You can adjust the settings in Windows by clicking **File | Options | Save | Save AutoRecover information every __ minutes**. In the Mac OS, click **Word | Preferences** and then **Output and Sharing | Save**. Enter a time for **Save every __ minutes**. However, when the content and format are complicated, it's a good idea to save the file manually from time to time by clicking 🖫 on the **Quick Access Toolbar** (see Figure A1.1).

Backing up your files

It should be obvious that your backup files should not be saved on your computer's hard drive. Table A1.1 lists some offline and online backup options along with their advantages and disadvantages. In terms of offline options, USB flash drives are probably your best bet for compactness and convenience, but external hard drives come with software that lets you schedule automatic backups. With an account and an internet connection,

TABLE A1.1 Possible options for backing up your files			
Backup Method	Capacity	Benefits	Drawbacks
USB flash drive	Up to 2 TB	Portable Can be encrypted No internet connection needed	Expensive per GB compared to hard drives Data lost if device is corrupted, damaged, or lost
SD/microSD card	Up to 512 GB	Portable Can be read from mobile devices	No internet connection needed Expensive per GB compared to hard drives Data lost if device is corrupted, damaged, or lost
External hard drive	Up to 20 TB	Portable Inexpensive Can be encrypted No internet connection needed Can perform automatic backups	Larger physical size compared to flash drives or SD cards Data lost if device is corrupted, damaged, or lost
Internal server with hard disk drives	Depends on organization	No subscription needed Files can be shared or kept private Can perform automatic backups	In-house networking required Internet connection required to access files remotely

you can take advantage of any number of online options. These include cloud services that store your files virtually, saving files to your organization's server, and even emailing files to yourself.

The most important thing about backing up your files is to have a plan. The odds are, unfortunately, that you will have a computer malfunction at least once every 3 years. For that reason alone:

- Schedule regular, automatic backups for your hard drive (that way you won't forget).
- Have a second backup option for important files that you are currently working on (especially projects with tight deadlines).
- Make sure your backup options don't run out of storage space. As a rough guide, a 10-page Word document takes up about 0.2 MB. In comparison, a picture taken with your smartphone can be as large as 12 MB, and downloaded videos take up hundreds of MB, depending on their resolution and length.

The takeaway message is: Protect your valuable files with a reliable backup system.

TABLE A1.1 *(continued)*

Backup Method	Capacity	Benefits	Drawbacks
Internal server with solid state drives	Depends on organization	Fastest bulk data storage No subscription needed Files can be shared or kept private Can perform automatic backups	Expensive In-house networking required Internet connection required to access files remotely
Cloud storage Google Drive Microsoft OneDrive Apple iCloud Amazon Drive Dropbox Box SugarSync	Infinite with subscription	2–50 GB of free storage Can perform automatic backups Can be accessed from any computer or mobile device with internet Files can be shared with collaborators	Subscription needed for larger storage tiers Prices increase with storage needs Internet connection required
Send emails to yourself	Typically 15 GB, potentially unlimited with organization email	Free Can be accessed from any computer with internet	Poor file organization Internet connection required 10-25 MB file size limit

Source: B. Knisely, personal communication, 14 Aug 2020

A1.3 AutoCorrect

AutoCorrect is useful for more than just correcting spelling mistakes. Increase your efficiency by programming AutoCorrect with a simple keystroke combination to replace an expression that takes a long time to type. Add a backslash (\) ahead of your abbreviated term to differentiate it from an existing acronym.

Long words

Let's say "beta-galactosidase" is a word you have to type frequently. Choose "\bg" to designate "beta-galactosidase." To program AutoCorrect for this entry:

Mac

1. Type "beta-galactosidase" and select it.
2. On the menu bar, click **Tools | AutoCorrect Options** or **Word | Preferences | Authoring and Proofing Tools | AutoCorrect.**
3. Type "\bg" in the **Replace** text box. You'll see that "beta-galactosidase" is automatically entered in the **With** text box.
4. Click the **Add** button.
5. Click the **red X** button to exit.
6. Next time you have to write "beta-galactosidase," simply type "\bg."

Windows

1. Type "beta-galactosidase" and select it.
2. Select **File | Options | Proofing | AutoCorrect Options**.
3. Type "\bg" in the **Replace** text box. You'll see that "beta-galactosidase" is automatically entered in the **With** text box.
4. Click the **Add** button.
5. Click **OK** to exit the **AutoCorrect** dialog box.
6. Click **OK** to exit the **Word Options** box.
7. Next time you have to write "beta-galactosidase," simply type "\bg."

Expressions with sub- or superscripts

Mac

1. First type the expression without sub- or superscripts (e.g., Vmax).
2. Select the characters to be sub- or superscripted, and then format them by clicking the appropriate button on the **Home** tab of the Ribbon (the expression then becomes V_{max}).
3. Select the entire expression and click **Tools | AutoCorrect Options**.
4. On the **AutoCorrect** tab, you will notice "Vmax" already entered in the **With** text box. Click the **Formatted text** option button to add the subscripting. The expression then becomes V_{max}.
5. Type "vmax" in the **Replace** text box.
6. Click the **Add** button.
7. Click the **red X** button to exit.

Windows

1. First type the expression without sub- or superscripts (e.g., Vmax).
2. Select the characters to be sub- or superscripted, and then format them by clicking the appropriate button on **Home | Font** (the expression then becomes V_{max}).
3. Select the entire expression and click **File | Options | Proofing | AutoCorrect Options**.
4. On the **AutoCorrect** tab, you will notice "Vmax" already entered in the **With** text box. Click the **Formatted text** option button to add the subscripting. The expression then becomes V_{max}.
5. Type "vmax" in the **Replace** text box.
6. Click the **Add** button.
7. Click **OK** to exit the **AutoCorrect** dialog box.
8. Click **OK** to exit the **Word Options** box.

Italicize scientific names of organisms automatically

By convention, scientific names of organisms are always italicized. When you type the name of the organism, capitalize genus but write species in lower case (e.g., *Homo sapiens*). To automate italicization, follow these steps:

Mac

1. First type the scientific name of the organism in your text (e.g., Aphanizomenon flos-aquae).
2. Italicize the name by selecting it and then clicking the **Italic** button (**I**) on the **Home** tab. The text then becomes: *Aphanizomenon flos-aquae.*

3. With the name still selected, click **Tools | AutoCorrect Options**.

4. On the **AutoCorrect** tab, click the **Formatted text** option button to display the italicized *Aphanizomenon flos-aquae* in the **With** text box.

5. Enter a shorter version of the name in the **Replace** text box, for example "aphan."

6. Click the **Add** button.

7. Click the **red X** button to exit.

Windows

1. First type the scientific name of the organism in your text (e.g., Aphanizomenon flos-aquae).

2. Italicize the name by selecting it and then clicking the **Italic** button (**I**) on **Home | Font**. The text then becomes: *Aphanizomenon flos-aquae*.

3. With the name still selected, click **File | Options | Proofing | AutoCorrect Options**.

4. On the **AutoCorrect** tab, you will notice Aphanizomenon flos-aquae (without italics) already entered in the **With** text box. Click the **Formatted text** option button to italicize it.

5. Enter a shorter version of the name in the Replace text box, for example "aphan."

6. Click the **Add** button.

7. Click **OK** to exit the AutoCorrect dialog box.

8. Click **OK** to exit the Word Options box.

A1.4 Endnotes in Citation-Sequence System

The most useful feature on the **References** tab, at least with regard to scientific papers, is the **Insert Endnote** (not footnote!) command for in-text references in the Citation-Sequence (C-S) system. To insert superscripted endnotes, type the sentence containing information that requires a reference. End the sentence with a period.

Mac

1. Then click **References | Insert Endnote**.

2. Type the full reference in proper format (see "The Citation-Sequence System" in Section 4.3).

3. Click **References | Show Notes** to exit the endnote and return to the text.

4. To change the endnote number style from Roman to Arabic, click **Insert | Footnote** on the menu bar (not the Ribbon) and select Arabic number format from the drop-down menu. Then click **Apply**.

Windows

1. Then click **References | Footnotes | Insert Endnote**.
2. Type the full reference in proper format (see "The Citation-Sequence System" in Section 4.3).
3. Click **References | Footnotes | Show Notes** to exit the endnote and return to the text.
4. To change the endnote number style from Roman to Arabic, click the Footnotes **dialog box launcher** (the diagonal arrow in the bottom right corner of the group) and select Arabic number format from the drop-down menu.

Word updates endnotes automatically so that they appear sequentially. To see an endnote, point the mouse at the superscripted number. To edit an endnote, double-click the number. To delete an endnote, highlight the superscripted number and press **Delete**. The other endnote numbers will be renumbered automatically.

A1.5 Equations

Mac

To insert a common mathematical equation or make your own, click **Insert | Equation** to display the **Equation** tab. Scan the built-in equations in the drop-down menu for the **Equation** button on the left side of the tab. If this menu does not contain the equation you'd like to write, choose a form to modify from the **Structures** group. Two commonly used structures are **Fraction** (numerator over denominator terms) and **Script** (subscripts and superscripts). Fill in numbers or select Greek letters, arrows, and mathematical operators from the **Math Symbols** group.

Windows

To insert a common mathematical equation or make your own, click **Insert | Equation** to display the **Equation Tools Design** tab. Scan the built-in equations in the drop-down menu for the **Equation** button on the left side of the tab. If this menu does not contain the equation you'd like to write, choose a form to modify from the **Structures** group. Two commonly used structures are **Fraction** (numerator over denominator terms) and **Script** (subscripts and superscripts). Fill in numbers or select Greek letters, arrows, and mathematical operators from the 8 symbol sets that appear after clicking the **More** arrow.

Note: If the Equation Editor that came with your version of Office 2019 doesn't work, Microsoft recommends downloading MathType software. Please visit the MS Support website for more information.

A1.6 Feedback Using Comments and Track Changes

Comments and Track Changes are editing tools that facilitate remote collaboration. You can send your paper to a reviewer in electronic format as an attached file, and your reviewer can send it back to you after making comments or revisions directly in the document.

Comments are frequently used to exchange ideas or ask questions when collaborating on a paper. You are most likely to add a comment, read someone else's comment and respond with a comment of your own, or delete one or more comments.

Some reviewers will make edits directly in the text using Track Changes. When this command is turned on, any changes made in the document will be colored and underlined. Each reviewer gets his or her own color so you can identify the changes made by multiple reviewers. In addition, a vertical line in the left margin alerts you to changes in your document. Clicking this line toggles between **All Markup** and **Simple Markup** modes.

Mac

Add a comment

1. Select the text you want to comment on.

2. Click **Review | New Comment**.

3. Type your comment in the box. Click outside of the comment area to return to the document.

Review and then delete comments   When there are no comments in a document, the **Delete**, **Previous**, and **Next** buttons are grayed out. When there are comments, these commands become available. After you have read and, if appropriate, taken action on all of the comments, delete them. To delete comments one at a time, click the comment balloon and then click **Delete**. Use the **Next** button in the Comment group to go to the next comment. To delete all of the comments at once, click the down arrow next to **Delete** and then **Delete All Comments in Document**.

Track changes

1. To turn on tracking, click **Review | Track Changes**. The button is green when tracking is turned on. To turn off tracking, click the button to display **Off**.

2. To review the edits made by others, click the **Next** button in the Reviewing group. Use the buttons for **Accept** or **Reject** or simply click **Next** to read the suggestion without taking any action.

3. Before you turn in your lab report, make sure all edits have been addressed (by accepting or rejecting them) and tracking is turned off. *Please note that accepting/rejecting changes does not remove the comments.* Comments must be deleted separately, as described in the previous section.

Windows

Add a comment

1. Select the text you want to comment on.

2. Click **Review | Comments | New Comment**.

3. Type your comment in the box. Click outside of the comment area to return to the document.

Review and then delete comments   When there are no comments in a document, the **Delete**, **Previous**, and **Next** buttons in the Comments group are not available. When there are comments, these commands become available. After you have read and, if appropriate, taken action on all of the comments, delete them with **Review | Comments | Delete | Delete All Comments in Document**. To delete comments one at a time, click the balloon and select **Delete Comment**.

Track changes

1. To turn on tracking, click **Review | Tracking | Track Changes**. The button is gray when tracking is turned on. To turn off tracking, click the button so that it turns white.

2. To review the edits made by others, click the **Next** button in the Changes group. Use the buttons for **Accept** or **Reject** or simply click **Next** to read the suggestion without taking any action.

3. *Please note that accepting/rejecting changes does not remove the comments.* Comments must be deleted separately, as described in the previous section.

Document inspector   Before you turn in your lab report, check that you've removed all comments and tracked changes. Click **File | Info | Check for Issues | Inspect Document**. Rather than selecting **Remove All**, click **Close**. Find the stray revision marks by clicking **Review | Changes | Next** and take action, if necessary, and delete or reject them.

A1.7 Format Painter

This paintbrush is great for copying and pasting format, whether it is the typeface or font size of a character or word, the indentation of a paragraph, or the position of tab stops. It's also handy for resuming numbering in a numbered list. It works with character, word, and paragraph formatting, but not page formatting.

To copy the format of a word, select the word, and click **Home | Paintbrush** (Mac) or **Home | Clipboard | Format Painter** (Windows). Drag the cursor (which has changed to a paintbrush next to an I-beam) over the text you want to format. To copy paragraph formatting (including tab stops), click inside the paragraph, but don't select anything. Then click the paintbrush and drag the mouse over the paragraph to which you want to apply the format.

To apply the format to multiple blocks of text, click the formatted text and then double-click the paintbrush (**Format Painter**). Apply the formatting to the selected blocks of text. Click the paintbrush again to turn off this function.

A1.8 Formatting Documents

An electronic file called "Biology Lab Report Template," available at **macmillanlearning.com/knisely6e**, is formatted according to these specifications.

Margins

Set 1.25" left and right and 1" top and bottom margins to give your instructor room to make comments.

Mac   Click **Layout | Margins**.

Windows   Click **Layout | Page Setup | Margins**.

Paragraphs

Mac   The **Paragraph** dialog box is opened by clicking **Format | Paragraph** on the menu bar.

First line indent is used to denote the beginning of a new paragraph. Under **Indentation | Special**, select **First line** from the drop-down menu and then specify **By: 0.5"** (or your preference).

Hanging indent is used for listing full references. The first line of each reference begins on the left margin and the subsequent lines are indented. First type each reference so that it is aligned on the left margin and ends with a hard return ¶. When you are finished, select all of the references. Under **Indentation | Special**, select **Hanging** from the drop-down menu and then specify **By: 0.25"** (or your preference).

Line spacing for all lines of text is adjusted with the **Line spacing** button. The default spacing is **Single**; your instructor may ask you to change the spacing to **Double** in order to have a little extra space to write comments.

Spacing between paragraphs. As an alternative to indenting the first line, paragraphs can be separated with a blank line. To automatically add extra space after each paragraph, thereby saving you the trouble of pressing the **Return** key twice, put the insertion pointer anywhere in the paragraph. In the **Paragraph** dialog box, enter **12 pt** in the **Spacing | After** box. This spacing corresponds to one blank line when using a 12 pt font size for your paper.

Line and page breaks. This feature is handy for preventing section headings from being separated from the body due to a natural page break. Select the section heading (for example, "Materials and Methods"). Then open the **Paragraph** dialog box and click the second tab called **Line and Page Breaks**. Under **Pagination**, check the box next to **Keep with next** to keep the heading together with the body. To keep all the lines of a paragraph together on the same page, check **Keep lines together**.

Windows   The **Paragraph** dialog box can be accessed from both the **Home** and **Layout** tabs. Clicking the diagonal arrow on the **Paragraph** group label launches a dialog box where you can format the following paragraph attributes.

First line indent is used to denote the beginning of a new paragraph. Next to **Indentation | Special**, select **First line** from the drop-down menu and then specify **By: 0.5"** (or your preference). If you use this method, change the default paragraph spacing to 0. See "Paragraph spacing" below.

Hanging indent is used for listing full references. The first line of each reference begins on the left margin and the subsequent lines are indented. First type each reference so that it is aligned on the left margin and ends with a hard return ¶. When you are finished, select all of the references and click **Home | Paragraph** dialog box launcher. Next to **Indentation | Special**, select **Hanging** from the drop-down menu and then specify **By: 0.25"** (or your preference).

Line spacing for all lines of text. The default spacing in Word 2019 is **Multiple at 1.08 lines**; your instructor may ask you to change the spacing to **Double** in order to have a little extra space to write comments. There is also a **Line spacing** command button under **Home | Paragraph** where you can make this change.

Paragraph spacing. As an alternative to indenting the first line, paragraphs can be separated with a blank line. The default after-paragraph spacing in Word 2019 is 8 pt. To increase this space to the height of a blank line, enter **12 pt** in the **Spacing | After** box. If you like the space between paragraphs, then it is not necessary to use the first line indent method.

Line and page breaks. This feature is handy for preventing section headings from being separated from the body due to a natural page break. Select the section heading (for example, "Materials and Methods"). Then open the **Paragraph** dialog box and click the second tab called **Line and Page Breaks**. Under **Pagination**, check the box next to **Keep with next** to keep the heading together with the body. To keep all the lines of a paragraph together on the same page, check **Keep lines together**.

Page numbers

Page numbers make it easier for you to assemble the pages of your document in the correct order.

Mac

1. To insert a page number, click **Insert | Page Number** from either the menu bar or the Ribbon, and then select a position for the numbers.

2. If you do not want the number to appear on the first page, uncheck **Show number on first page**.

3. To start numbering with a custom number (for example, if the documents are chapters in a book), click **Insert | Page Number**, and enter a number under **Page numbering | Start at: __**.

Windows

1. To insert a page number, click **Insert | Header & Footer | Page Number**, and then select a position for the numbers.

2. If you do not want the number to appear on the first page, click **Different first page**.

3. To start numbering with a custom number (for example, if the documents are chapters in a book), click **Insert | Header & Footer | Page Number | Format Page Numbers**, and enter a number under **Page numbering | Start at: __**.

A1.9 Proofreading Your Documents

Before you waste reams of paper printing out drafts of your document, have Word check spelling and grammar. Use your eyes to look over format on-screen before proofreading the printed document.

Spelling and grammar

Word gives you visual indicators to alert you to possible spelling and grammar mistakes. Words that are possibly misspelled are underlined with a wavy red line. Phrases that may contain a grammatical error or an extra space are underlined with a double blue line in both the Mac and Windows operating systems. Do not ignore these visual cues!

Mac   To deal with a word underlined in red, **control-click** it. A pop-up menu appears with commands and suggestions for replacements. **Ignore All** applies only to the current document. **Add to Dictionary** applies to all future documents. It makes sense to add scientific terminology to Word's dictionary after you consult your textbook or laboratory manual to confirm the correct spelling. After making a selection on the pop-up menu, the wavy red underline is deleted and the word is ignored in the manual, systematic spelling and grammar check. Similarly, to deal with a possible grammatical error underlined in blue (including extra spaces between words), **control-click** it to accept or ignore Word's suggestions.

Windows   To deal with expressions underlined in red or blue, right-click them. A pop-up menu appears with commands and suggestions for replacements. The options are the same as for Mac above.

Prevent section headings from separating from their body

Research articles are divided into sections: Abstract, Introduction, Materials and Methods, Results, Discussion, and References. Each section begins with a heading, on a separate line, followed by the body. When you check the format of your paper, make sure the heading is not cut off from the body of the section.

To prevent heading-body separation problems, use one of these options:

- See "Line and page breaks" in the Paragraphs section.
- Insert a hard page break to the left of the heading to force the heading onto the next page with the body. On a Mac, press ⌘ **(command)+Return**. In Windows, press **Ctrl+Enter**.

Prevent figures and tables from separating from their caption

See "Prevent section headings from separating from their body."

Errant blank pages

To delete blank pages in the middle of a document, go to the blank page, click it, and display the hidden symbols. For Mac, click **Home | Show/**

Hide ¶. For Windows, click **Home | Paragraph | Show/Hide** ¶. Delete spaces and paragraph symbols until the page is gone.

Similarly, if the blank page is at the end, navigate to the end of the document and remove the hidden symbols to delete the extra page.

Finally, print a hard copy

When you are confident that you've found and corrected all mistakes on-screen, print a hardcopy and proofread your paper again. Some mistakes are more easily identified on paper, and it is always better for you, rather than your instructor, to find them.

A1.10 Sub- and Superscripts

Expressions with sub- or superscripting are common in the natural sciences and their formatting makes them readily identifiable to members of the scientific community. Therefore, it would be incorrect to write sub- or superscripted characters on the same line as the rest of the text. Similarly, when using scientific notation, exponents are always superscripted. It is not acceptable to use a caret (^) to designate superscript or an uppercase E to represent an exponent.

> **RIGHT:** 2×10^{-3} (exponent is superscripted)

> **WRONG:** $2 \times 10{-}3$ or $2 \times 10\text{^}{-}3$ or $2 \times 10\text{E}{-}3$

If you frequently have to type expressions with sub- and superscripts, save time by programming them in AutoCorrect (see Section A1.3).

Mac

To superscript or subscript text, first type the expression without formatting. Then highlight the character(s) to be sub- or superscripted and click the appropriate button on the **Home** tab of the Ribbon. Alternatively, you can apply and remove sub- and superscripting using keyboard commands ("hot keys"). Select the character(s), and then hold down ⌘ **(command)** while pressing = for subscript or hold down both ⌘ and **Shift** while pressing = for superscript.

Windows

To superscript or subscript text, first type the expression without formatting. Then highlight the character(s) to be sub- or superscripted and click the appropriate button under **Home | Font**. Alternatively, you can apply and remove sub- and superscripting using keyboard commands ("hot keys").

Select the character(s), and then hold down **Ctrl** while pressing = for subscript or hold down both **Ctrl** and **Shift** while pressing = for superscript.

A1.11 Symbols

Mac

AutoCorrect for symbols   All of the Greek letters and many mathematical symbols have already been defined in Math AutoCorrect. To see the full list, click **Tools | AutoCorrect Options** and then select the **Math AutoCorrect** tab. Most of the entries begin with a backslash. So, for example, if you wanted to insert a degree symbol in your spreadsheet, type \degree, and Word automatically converts this to °. This feature will only work if the **Use Math Auto-Correct rules outside of math regions** checkbox on the **Math AutoCorrect** tab is checked.

For symbols not on this list, click **Insert | Advanced Symbol** and scroll through the collection. Once you find the symbol you need, you can either add the symbol to the AutoCorrect list or make a keyboard shortcut. These options are explained below.

Add symbols to AutoCorrect   Let's use $\bar{x}$ (xbar), the mean of a set of observations, as an example.

1. Copy and paste $\bar{x}$ from the internet.
2. Now select the newly created character, and click **Tools | AutoCorrect** on the menu bar. It doesn't matter if you add the new symbol to AutoCorrect or Math AutoCorrect.
3. The $\bar{x}$ symbol will already be entered in the **With** text box and the **Formatted text** option box will be selected. Type a unique combination of characters in the **Replace** text box, for example, \xbar. Click the **Add** button, and click the **red X** button to exit.
4. Next time you want to insert the $\bar{x}$ character, simply type "\xbar".

Define hot keys for symbols   For frequently used symbols, you can also make a keyboard shortcut in the Symbols dialog box. Let's use the degree sign as an example.

1. After clicking this symbol in the **Symbol** dialog box (Unicode character F0B0), click the **Keyboard Shortcut** button at the bottom of the box.
2. In the **Customize Keyboard** dialog box, define a combination of keystrokes using **Control**, ⌘, **Control+Shift**, or **Control+⌘** plus some other character. Because the degree sign looks like a lowercase *o*, let's use **Control+o**. To define **Control+o** as the keyboard

shortcut for the degree sign, hold down **Control** while pressing **o** in the **Press new shortcut key** box.

3. Word notifies you that this combination is unassigned or that it has already been assigned to another command. (Note: Even though you typed **Control** and lowercase **o**, the shortcut appears as Control and uppercase O in the box. If you had typed **Control** and uppercase **O**, this would have appeared as Control+Shift+O.)

4. Next time you have to write the symbols for "degrees Celsius," simply type **Control+o** followed by uppercase **C** to get: °C.

If the shortcut has already been assigned to a different command, assigning a symbol will automatically overwrite the original command. In other words, the same keystroke combination will insert the symbol instead of carrying out the original command. Do not overwrite the original command if this is one you use often!

Windows

AutoCorrect for symbols   All of the Greek letters and many mathematical symbols have already been defined in Math AutoCorrect. To see the full list, click **File | Options | Proofing | AutoCorrect options** and then select the **Math AutoCorrect** tab. Most of the entries begin with a backslash. So, for example, if you wanted to insert a degree symbol in your spreadsheet, type \degree, and Word automatically converts this to °. This feature will only work if the **Use Math AutoCorrect rules outside of math regions** checkbox on the **Math AutoCorrect** tab is checked.

For symbols not on this list, click **Insert | Symbols | Symbol | More Symbols** and scroll through the collection. Once you find the symbol you need, you can either add the symbol to the AutoCorrect list or make a keyboard shortcut. These options are explained below.

Add symbols to AutoCorrect   Let's use $\bar{x}$ (xbar), the mean of a set of observations, as an example.

1. To make $\bar{x}$ initially, type x and then click **Insert | Symbol | More Symbols | Subset: Combining Diacritical Marks**. Click the "combining overline" (Unicode character 0305).

2. Now select the newly created character, and click **File | Options | Proofing | AutoCorrect options** on the Ribbon. It doesn't matter if you add the new symbol to AutoCorrect or Math AutoCorrect.

3. The $\bar{x}$ symbol will already be entered in the **With** text box and the **Formatted text** option box will be selected. Type a unique combination of characters in the **Replace** text box, for example, \xbar.

Click the **Add** button, click **OK** to exit the **AutoCorrect** dialog box, and finally click **Close** to exit the **Symbol** box.

4. Next time you want to insert the $\bar{x}$ character, simply type "\xbar".

Define hot keys for symbols   For frequently used symbols, you can also make a keyboard shortcut in the Symbols dialog box. Let's use the degree sign as an example.

1. After clicking this symbol in the Symbol dialog box (Unicode character F0B0), click the **Keyboard Shortcut** button at the bottom of the box.

2. In the Customize Keyboard dialog box, define a combination of keystrokes using **Ctrl**, **Alt**, **Ctrl+Shift**, or **Ctrl+Alt** plus some other character. Because the degree sign looks like a lowercase o, let's use **Control+o**. To define **Control+o** as the keyboard shortcut for the degree sign, hold down **Control** while pressing **o** in the **Press new shortcut key** box.

3. Word notifies you that this combination is unassigned or that it has already been assigned to another command. (Note: Even though you typed **Control** and lowercase **o**, the shortcut appears as Control and uppercase O in the box. If you had typed **Control** and uppercase **O**, this would have appeared as Control+Shift+O.)

4. Next time you have to write the symbols for "degrees Celsius," simply type **Control+o** followed by uppercase **C** to get: °C.

If the shortcut has already been assigned to a different command, assigning a symbol will automatically overwrite the original command. In other words, the same keystroke combination will insert the symbol instead of carrying out the original command. Do not overwrite the original command if this is one you use often!

A1.12 Tables

Mac

Creating a table

1. Position the insertion pointer where you want to insert the table.

2. Click **Insert | Table**. Highlight the desired number of columns and rows. A blank table appears with the cursor in the first cell. To change the table format, use the **Table Design** and **Table Layout** tabs on the Ribbon. These tabs only appear when the insertion pointer is inside the table.

3. To insert or delete columns or rows or to merge or split cells, use the command buttons on the **Table Layout** tab.

Formatting text within tables
1. To apply formatting to adjacent cells, first select the cells by clicking the first cell in the range, holding down the **Shift** key, and then clicking the last cell.
2. With the cells still selected, apply, for example, boldface to the text by clicking **Home | B (Bold)**.

Viewing gridlines   It is easier to enter data in a table when the gridlines are displayed. Position the insertion pointer inside the table and click **Table Layout | View Gridlines**. Gridlines are not printed.

Lines in a table   By convention, tables in scientific papers do not have vertical lines to separate the columns, and horizontal lines are used only to separate the table caption from the column headings, the headings from the data, and the data from any footnotes (notice the format of the tables in this book).

1. Select all of the cells. Click **Table Design | Borders** and select **No Border**. You will still see the gridlines if they are selected (see "Viewing gridlines"), but gridlines are not printed.
2. Select the cells in the first row. Click **Table Design | Borders** and check **Bottom Border** and **Top Border**.
3. Select the cells in the last row. Click **Table Design | Borders** and check **Bottom Border**.

Navigating in tables
1. To jump from one cell to an adjacent one, use the arrow keys.
2. To move forward across the row, use the **Tab** key. Note: If you press **Tab** when you are in the last cell of the table, Word adds another row to the table.
3. To align text on a tab stop in a table cell, press **Ctrl+Tab**.

Anchoring images within tables   Getting images to line up horizontally and vertically on a page can be tricky. It's possible to make multiple columns using **Layout | Columns** or to adjust the position of each picture on the **Picture Format** tab, but quite often the images become misaligned when the document is edited. To make images stay put, create a table. Select the desired layout (for example, 3 columns × 1 row), remove all borders, size the images so they fit into the cells, and then insert the images into the table.

Repeating header rows   When a table extends over multiple pages, it is convenient to have Word automatically repeat the header row at the top of each page. To activate this command, click anywhere in the header row and then select **Table Layout | Repeat Header Rows**.

Windows

Creating a table

1. Position the insertion pointer where you want to insert the table.
2. Click **Insert | Tables | Table**. Highlight the desired number of columns and rows.

A blank table appears with the cursor in the first cell. To change the table format, use the **Table Tools Design** and **Table Tools Layout** tabs. These tabs only appear when the insertion pointer is inside the table.

To insert or delete columns or rows or to merge or split cells, use the command buttons on the **Table Tools Layout** tab.

Formatting text within tables

1. To apply formatting to adjacent cells, first select the cells by clicking the first cell in the range, holding down the **Shift** key, and then clicking the last cell.
2. With the cells still selected, apply, for example, boldface to the text by clicking **Home | Font | B (Bold)**.

Viewing gridlines   It is easier to enter data in a table when the gridlines are displayed. Position the insertion pointer inside the table and click **Table Tools Layout | Table | View Gridlines**. Gridlines are not printed.

Lines in a table   By convention, tables in scientific papers do not have vertical lines to separate the columns, and horizontal lines are used only to separate the table caption from the column headings, the headings from the data, and the data from any footnotes (notice the format of the tables in this book).

1. Select all of the cells. Click **Table Tools Design | Borders | Borders** and select **No Border**. You will still see the gridlines if they are selected (see "Viewing gridlines"), but gridlines are not printed.
2. Select the cells in the first row. Click **Table Tools Design | Borders | Borders** and check **Bottom Border** and **Top Border**.

3. Select the cells in the last row. Click **Table Tools Design | Borders | Borders** and check **Bottom Border**.

Navigating in tables

1. To jump from one cell to an adjacent one, use the arrow keys.
2. To move forward across the row, use the **Tab** key. Note: If you press **Tab** when you are in the last cell of the table, Word adds another row to the table.
3. To align text on a tab stop in a table cell, press **Ctrl+Tab**.

Anchoring images within tables   Getting images to line up horizontally and vertically on a page can be tricky. It's possible to make multiple columns using **Layout | Page Setup | Columns** or to adjust the position of each picture on the **Picture Tools Format** tab, but quite often the images become misaligned when the document is edited. To make images stay put, create a table. Select the desired layout (for example, 3 columns × 1 row), remove all borders, size the images so they fit into the cells, and then paste the images into the table.

Repeating header rows   When a table extends over multiple pages, it is convenient to have Word automatically repeat the header row at the top of each page. To activate this command, click anywhere in the header row and then select **Table Tools Layout | Data | Repeat Header Rows**.

Go to **macmillanlearning.com/knisely6e**
and select "Student Site" to access
samples, template files, and tutorial videos

Making Graphs in Microsoft Excel

A2.1 Introduction

Spreadsheets are a common and convenient way to collect and share data. Google Sheets is often used for this purpose because it is free, it requires no special software, it can be used on any computer (Windows or Mac), it allows real-time collaboration (multiple users can edit the document at the same time), and the files can be downloaded in Microsoft Excel format.

The main disadvantage to Google Sheets is that this program lacks MS Excel's advanced features for doing calculations and graphing the data. To get the best of both worlds, collect data in Google Sheets, download the file as an Excel file, and then process and analyze the data in the *desktop* version of Excel. As explained in Section A1.1, Microsoft Office 365 supports online and desktop versions of the software, but the online version lacks some important features of the desktop version.

This appendix has three parts. The "Spreadsheet Management" section explains how to enter data, navigate within the spreadsheet, change the number of displayed decimal places, insert symbols, sort data, add and rename worksheets, and format the spreadsheet for printing. The "Formulas" section introduces you to using formulas and functions to save time on data analysis. The bulk of this appendix gives detailed instructions for Windows and Mac on how to make different types of graphs and format them according to Council of Science Editors (CSE) standards. See also the video tutorials for Windows and Mac at **macmillanlearning.com/knisely6e**. For all other Excel questions, please visit the Microsoft Office support website.

In this book, the nomenclature for the command sequence is as follows: **Ribbon tab | Group | Command button | Additional Commands** (if available).

For example, to make an XY graph in Windows, **Insert | Charts | Insert Scatter (X, Y) ▼ | Scatter with Straight Lines and Markers** means "Select the **Insert** tab on the Ribbon, and in the **Charts** group, click the down arrow on the **Insert Scatter (X, Y)** button, and select the

Scatter with Straight Lines and Markers option." In the Mac OS, the corresponding command sequence is **Insert | X Y (Scatter) ▼ | Scatter with Straight Lines and Markers.**

A2.2 Spreadsheet Management

The tasks are listed in alphabetical order.

Add worksheets to the same file

Each Excel workbook comes with one worksheet, named Sheet1 (Figure A2.1). You can give this sheet a different name by double-clicking the tab in the lower left corner of the screen. You can also add more worksheets by clicking the **+ New sheet** button in Windows or the corresponding **+ Insert Sheet** button in the Mac OS. Some good reasons for keeping multiple worksheets in one workbook are:

- To collect replicate data, yet keep individual trials separate.
- To collect data from multiple lab sections for the same experiment, to pool or keep separate as needed.
- To simplify file organization.

You can rearrange the worksheets by dragging a tab left or right. To the left of the sheet tabs, you'll see two sheet tab scroll buttons. The scroll buttons are only available if the workbook contains so many worksheets that their tabs cannot all be displayed at once.

Change view

The **View** tab in Excel gives you options for displaying your workbook(s) on the screen. There are three workbook views that can also be accessed from the status bar (see Figure A2.1).

- **Normal** view is the default view and is typically used when entering data.
- **Page Layout** view displays margins; headers, footers, and page numbers; and page breaks.
- **Page Break Preview** is handy for adjusting page breaks to keep blocks of text or data together.

Two other views, called **Freeze Panes** and **Split**, are useful to view row and column headings while scrolling through the spreadsheet. **Split** view splits the worksheet into 2 or 4 sections (based on the cell that is selected), each of which can be scrolled independently. Use **Split** view to view different sections of the worksheet simultaneously. The **Freeze Panes** command allows you to lock row and column headings while scrolling

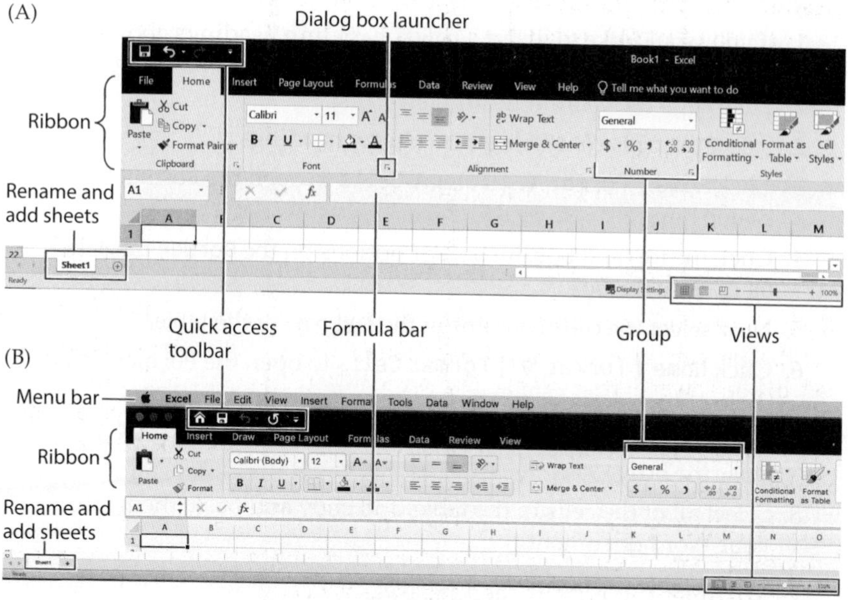

(A)

(B)

Figure A2.1   Screen display of part of the command area in (A) an Excel 2019 worksheet, and (B) an Excel for Mac 2019 worksheet. The status bar at the bottom of the screen has commands for changing views and adding and renaming worksheets.

through the rest of the worksheet. To lock the headings, click the cell just below the row and just to the right of the column that contains the headings. Then click one of the **Freeze** options.

Create tables

Excel worksheets are technically tables that have more than a million rows and more than 16,000 columns. Most of the time you will analyze data and make graphs in Excel without actually printing out the worksheet. However, your instructor may ask you to attach a table containing the raw data in an appendix. This is different from inserting a table in the lab report itself. It is *not necessary* to include a table when you already have a graph that shows the same data. Make *either* a table or a graph—not both—depending on your objectives (see Section 5.4).

By convention, tables in scientific papers do not have vertical lines to separate the columns, and horizontal lines are used only to separate the table caption from the column headings, the headings from the data, and the data from any footnotes. To format a table like this in Excel, follow these instructions.

Mac

1. If you would like to insert a row for column headings above the data, click a row and then **Home | Insert ▼ | Insert Sheet Rows**. Then type the column headings.

2. Select the cells containing the column headings.

3. Click **Home | Format ▼ | Format Cells** to open the Format Cells dialog box.

4. Click the top and bottom border options on the **Border** tab. Click **OK** to close the box.

5. Now select the cells containing the last row of the table.

6. Click **Home | Format ▼ | Format Cells** to open the Format Cells dialog box.

7. Click the bottom border option on the **Border** tab. Click **OK** to close the box.

8. Select all of the cells in the table, and copy and paste them into your Word document.

9. Type a table caption *above* the table.

Windows

1. If you would like to insert a row for column headings above the data, click a row and then **Home | Cells | Insert ▼ | Insert Sheet Rows**. Then type the column headings.

2. Select the cells containing the column headings.

3. Click **Home | Cells | Format ▼ | Format Cells** to open the Format Cells dialog box.

4. Click the top and bottom border options on the **Border** tab. Click **OK** to close the box.

5. Now select the cells containing the last row of the table.

6. Click **Home | Cells | Format ▼ | Format Cells** to open the Format Cells dialog box.

7. Click the bottom border option on the **Border** tab. Click **OK** to close the box.

8. Select all of the cells in the table, and copy and paste them into your Word document.

9. Type a table caption *above* the table.

Enter data

Organizing your data from the start will save you time in the long run. Save the top row and possibly the first column for headings. Make the headings unambiguous by clearly stating the variable along with the units. For reference later, you may also want to include a few comments about the experimental conditions or the samples, especially if you plan to collect multiple data sets in the same spreadsheet.

Enter data in adjacent columns and rows—don't skip columns to space out the data. This arrangement will make it easier to navigate large data sets and to plot the data later. The column width can be adjusted for long headings and large numbers by double-clicking the boundary between the column letters (A, B, C, D, etc.) or simply dragging the boundary to the desired width. Wrapping text is a way to prevent the columns from becoming excessively wide.

Non-numerical data (text) is left-aligned in Excel; numerical data is right-aligned. When entering numerical data, enter only the value without the units, otherwise formulas applied to the data will not work.

Even if all the data have not yet been entered, it's a good idea to save the spreadsheet file under a memorable name. "C-fern lab" or "lac operon lab" are searchable file names that provide far more useful information than "biolab."

Format cells

Home | Cells | Format | Format Cells opens a dialog box with 6 tabs: Number, Alignment, Font, Border, Fill, and Protection. For most cells containing numbers, **Number | General** works well. To limit the number of decimal places in the selected cells, click **Number | Number**, enter the desired number of decimal places, and then click **OK. Number | Scientific** is not recommended, because Excel's use of E to represent *exponent* is not accepted by the Council of Science Editors.

Navigate within a spreadsheet

The keyboard shortcuts in Table A2.1 allow you to move around a large data set quickly.

Print spreadsheet

The **Page Layout** tab has the buttons that allow you to format your spreadsheet if you need to print it out:

- Change the margins
- Add headers, footers, or page numbers
- Scale the content to fit on one sheet of paper
- Adjust page breaks
- Set the print area
- Repeat row or column headings on multiple-page worksheets

TABLE A2.1 Navigation shortcuts in Excel		
Navigate to	Mac	Windows
A1	Ctrl + Home	Ctrl + Home (desktop)
		Ctrl + Fn + Home (some laptops)
Last cell that contains data	Ctrl + End	Ctrl + End (desktop)
		Ctrl + Fn + End (some laptops)
Edges of the data set	Cmd + arrows	Ctrl + arrows

Round up final values

When a calculation results in a value that has more decimal places than the measurements from which it originated, round up the final value. In other words, make sure that your final answer does not have more decimal places than those in the original measurement.

> EXAMPLE: A digital spectrophotometer can read absorbance to the thousandths place, e.g. 0.235. Let's say that after Excel averages 10 replicate absorbance measurements, the result is 0.24158. The average absorbance should be reported as 0.242.

Only round up the *final* calculation result, not the intermediate values of a multi-step calculation, otherwise you will introduce error. To round up the final result, click the desired cell and then **Home | Number | Decrease Decimal** repeatedly until the appropriate number of decimal places is displayed.

Save and back up computer files

Good file organization, as described in Section A1.2, is just as important for Excel workbooks as Word documents. Review this section to develop good habits for naming, saving, and backing up computer files.

Sort data

If you'd like to alphabetize a list of names or arrange numerical values in ascending or descending order, use the **Home | Editing | Sort & Filter** command. First, select *all of the columns that contain data* that you want to sort. This is important, because if you select only one column, then only the data in that column will be sorted. Data in the adjacent columns will then no longer correspond to the correct data in the sorted column. After selecting all of the data, click **Custom Sort** on the **Sort & Filter** button's drop-down list. Select the criteria according to which you want to sort: by column, by values, and in which order. Then click **OK**.

Subscripts, superscripts, and symbols

Special characters are common in biology and other STEM fields. Their formatting makes them readily identifiable to members of the scientific community. You may need to type exponents or expressions with sub- and superscripts, Greek letters, and mathematical symbols in column headings and in the axis labels of your graphs.

First type the expression without sub- and superscripts. Then highlight the character(s) to be sub- or superscripted and click the appropriate button. In Windows, the buttons for sub- and superscripting are found on the **Home** tab in the **Font** group (see Figure A2.1). Click the **dialog box launcher**—the diagonal arrow—to access these buttons. On a Mac, the cell format for any cells containing sub- and superscripts must be changed to "Text" (see "Format cells"). Type the expression without formatting, highlight the character(s) to be sub-or superscripted, and click **Format** on the menu bar. Then click **Cell | Effects: Subscript** or **Superscript**.

The buttons for Greek letters and mathematical symbols are found under **Insert | Symbol** on the far right of the tab.

Wrap text

When you type a long text in a cell, it runs into the adjacent cell to the right. If there is text or a numerical entry in the adjacent cell, the long text is hidden (it appears to be cut off). To make the long text visible, you could widen the columns, but then fewer columns will be visible on the screen. To be able to maintain column width and still view the entire text, click the desired cell and then **Home | Alignment | Wrap Text**. Excel expands the row height to accommodate the contents of the cell.

A2.3 Formulas in Excel

Excel is a popular spreadsheet program in the business world, but its "number crunching" capabilities make it a powerful tool for data reduction and analysis in general. You can use Excel like a calculator by typing numbers and mathematical operators into a cell and then pressing **Enter**. Most likely, however, you will write your own formulas or choose common functions from Excel's collection and apply them to cell references instead of numbers. Excel is great for doing repetitive calculations quickly.

In this section you will learn how to write some formulas frequently used in biology. Even when you've become proficient at writing formulas in Excel, however, it's still a good idea to *do a sample calculation by hand (using your calculator) to make sure the formula you entered in Excel gives you the same result.* If you find a discrepancy, check the formula and check your math. Make sure the answer makes sense.

TABLE A2.2 Operators commonly used for calculations in Excel

Operator	Meaning
+	Addition
−	Subtraction or negation
*	Multiplication
/	Division
%	Percentage
^	Raised to the exponent
:	Range of adjacent cells
,	Multiple, non-adjacent cells

Operators

Some commonly used operators are shown in Table A2.2. Excel performs the calculations in order from left to right according to the same order of operations used in algebra: first negation (−), then all percentages (%), then all exponentiations (^), then all multiplications and divisions (* or /), and finally all subtractions and additions (− or +). For example, the formula "=100−50/10" would result in "95", because Excel performs division before subtraction. To change the order of operations, enclose part of the formula in parentheses. For example, "=(100−50)/10" would result in "5."

Cell references

The nomenclature for cell references is a letter, representing the column, and a number, representing the row; for example, A1. Cell references are *relative* by default, so that when a formula containing a cell reference is copied to another cell, the same operation will be carried out on different values. However, sometimes you might want a *specific* cell to be used in a formula, which does not change when the formula is copied to another cell. To make cell references *absolute*, insert a dollar sign ($) ahead of the letter and number.

The difference between relative and absolute cell references is shown in Table A2.3. First, notice that formulas in Excel always start with an equal sign (=) followed by the cell references, numbers, and operators that make up the formula. The original formula in this example was typed in cell D3. This formula takes the sum of the values in B3 and C3 and divides this sum by the sum of the values in B2 and C2. When this formula is copied to cells D4 and D5, you can see that Excel adjusts the cell references automatically; the results are shown in E3-E5.

TABLE A2.3 The difference between relative cell references and absolute cell references. Formulas that contain relative cell references automatically adjust for position when the formula is copied. To reference a *specific* cell in a formula, add a dollar sign ($) in front of both the column letter and the row number. In this example, to always use the values of the constants in B2 and C2, make these cell references absolute.

	A	B	C	D	E	F	G
1				Formula with relative cell references	Result with relative cell references	Formula with absolute cell references	Result with absolute cell references
2	Constants	100	200				
3	Expt 1	1	2	=(B3+C3)/ (B2+C2)	0.01	=(B3+C3)/ ($B$2+$C$2)	0.01
4	Expt 2	3	4	=(B4+C4)/ (B3+C3)	2.333333	=(B4+C4)/ ($B$2+$C$2)	0.023333
5	Expt 3	5	6	=(B5+C5)/ (B4+C4)	1.571429	=(B5+C5)/ ($B$2+$C$2)	0.036667

The cell references for the constants were made absolute in cell F3. When this formula was copied to cells F4 and F5, only the relative cell references changed; the results are shown in G3-G5.

Formula syntax

This section explains how to enter a formula in the correct format and to apply the formula to a range of cells.

Mac

TYPE A FORMULA IN THE ACTIVE CELL

1. Click a cell in which you want the result of the formula to be displayed (the so-called **active cell**). The selected cell will have a green border with a small square in the lower right corner, as in cell L4 in Figure A2.2B. The small square is called the **fill handle** and is used to copy formulas (see "Copying formulas using the fill handle").
2. Type "=" (equal sign).
3. Type the constants, operators, cell references, and functions that you want to use in the calculation. See Table A2.4 for examples.
4. Press **Return**. The result of the calculation is displayed in the active cell. You can view and edit the formula on the **formula bar**.

(A)

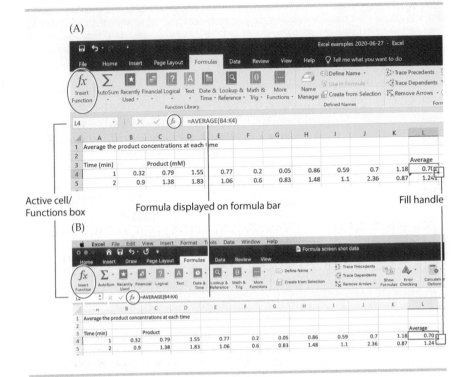

Active cell/
Functions box

Formula displayed on formula bar

Fill handle

(B)

Figure A2.2   The **Formulas** tab has an **Insert Function** button and other commands for doing calculations in (A) Excel 2019, and (B) Excel for Mac 2019.

APPLY A FUNCTION TO VALUES IN ADJACENT CELLS

1. Click a cell in which you want the result of the formula to be displayed.

2. Type "=", the name of the function (e.g., AVERAGE) and then "(".

3. Hold down the left mouse button and select the cells you want to average. If there are a lot of cells to average, click the first cell, hold down the **Shift** key, and then click the last cell in the range. Regardless of the selection method, Excel automatically inserts the first cell and last cell of the range, separated by a colon, in the formula bar (see Figure A2.2B).

4. Type ")" and press **Return**.

APPLY A FUNCTION TO VALUES NOT IN ADJACENT CELLS

1. Click a cell in which you want the result of the formula to be displayed.

2. Type "=", the name of the function (e.g., AVERAGE) and then "(".

TABLE A2.4 Examples of formulas written in Excel

Formula	Meaning
=16*31-42	Because multiplication is done before subtraction, the result is (16 x 31) minus 42.
=A1	The value in cell A1 is assigned to the active cell. If the value in A1 changes, the value will change automatically in all cells that reference A1.
=$A$1	The value in cell A1 becomes absolute. When an absolute cell reference is used in a formula, the value does not change when the formula is copied.
=A1+B1	The sum of the values in cells A1 and B1 is assigned to the active cell. Cell references eliminate the risk of introducing typos when retyping values.
=(O4/H4)^(10/(I4−B4))	The operations in parentheses are done first. The value in cell O4 is divided by the value in cell H4. This number is raised to the power of 10 divided by the difference between the values in cells I4 and B4.
=SUM(A1:A26)	The sum of the values in cells A1 through A26 inclusive is assigned to the active cell. To select a range of adjacent cells, use **Shift**.
=SUM(B4,J4,L4,T4)	The sum of the values in cells B4, J4, L4, and T4 is assigned to the active cell. To select a range of nonadjacent cells, use **Ctrl** (Windows) or ⌘ (Mac).
=AVERAGE(B4:U4)	The average of the values in cells B4 through U4 inclusive is assigned to the active cell. To select a range of adjacent cells, use **Shift**.
=AVERAGE(B4,J4,L4,T4)	The average of the values in cells B4, J4, L4, and T4 is assigned to the active cell. To select a range of nonadjacent cells, use **Ctrl** (Windows) or ⌘ (Mac).
=log(B1)	The log of the value in cell B1 is assigned to the active cell.
=SQRT(B1)	The square root of the value in cell B1 is assigned to the active cell.

3. Hold down the ⌘ (**Cmd**) key and click the cells you want to average. Excel automatically inserts the selected cells, separated by commas, in the formula bar.

4. Type ")" and press **Return**.

SELECT A STATS FUNCTION WITH THE ΣAUTOSUM BUTTON

1. Click a cell in which you want the result of the formula to be displayed.

2. Click **Home | ΣAutoSum ▼** or **Formulas | ΣAutoSum ▼** to sum, average, or count numbers, determine max and min, or choose

more functions (see Figure A2.2B). If necessary, adjust the range of cells to which the function will be applied.

SELECT AN EXCEL FUNCTION FROM THE ACTIVE CELL/FUNCTIONS LIST

1. Click a cell in which you want the result of the formula to be displayed.
2. Type "=" in the active cell. The most recently used function is displayed in the **Active Cell/Functions** box to the left of the formula bar (see Figure A2.2B).
3. Click the down arrow to select a function from the drop-down menu. The Formula Builder pane explains the function you've selected.
4. On the formula bar, enter the range of cells to which the function will be applied.

Windows

TYPE A FORMULA IN THE ACTIVE CELL

1. Click a cell in which you want the result of the formula to be displayed (the so-called **active cell**). The selected cell will have a green border with a small square in the lower right corner, as in Figure A2.2A. The small square is called the **fill handle** and is used to copy formulas (see "Copying formulas using the fill handle").
2. Type "=" (equal sign).
3. Type the constants, operators, cell references, and functions that you want to use in the calculation. See Table A2.4 for examples.
4. Press **Enter**. The result of the calculation is displayed in the active cell. You can view and edit the formula on the **formula bar**.

APPLY A FUNCTION TO VALUES IN ADJACENT CELLS

1. Click a cell in which you want the result of the formula to be displayed.
2. Type "=", the name of the function (e.g., AVERAGE) and then "(".
3. Hold down the left mouse button and select the cells you want to average. If there are a lot of cells to average, click the first cell, hold down the **Shift** key, and then click the last cell in the range. Regardless of the selection method, Excel automatically inserts the first cell and last cell of the range, separated by a colon, on the formula bar (see Figure A2.2A).
4. Type ")" and press **Enter**.

APPLY A FUNCTION TO VALUES NOT IN ADJACENT CELLS

1. Click a cell in which you want the result of the formula to be displayed.
2. Type "=", the name of the function (e.g., AVERAGE) and then "(".
3. Hold down the **Ctrl** key and click the cells you want to average. Excel automatically inserts the selected cells, separated by commas, on the formula bar.
4. Type ")" and press **Enter**.

SELECT A STATS FUNCTION WITH THE ΣAUTOSUM BUTTON

1. Click a cell in which you want the result of the formula to be displayed.
2. Click **Home | Editing | ΣAutoSum ▼** or **Formulas | Function Library | ΣAutoSum ▼** to sum, average, or count numbers, determine max and min, or choose more functions (see Figure A2.2A). If necessary, adjust the range of cells to which the function will be applied.
3. Click **OK** or press **Enter**.

SELECT AN EXCEL FUNCTION FROM THE
ACTIVE CELL/FUNCTIONS LIST

1. Click a cell in which you want the result of the formula to be displayed.
2. Type "=" in the active cell. The most recently used function is displayed in the **Active Cell/Functions** box to the left of the formula bar (see Figure A2.2A).
3. Click the down arrow to select a function from the drop-down menu. The **Function Arguments** dialog box appears.
4. Select the range of cells for the function and click **OK**.

Copying formulas using the fill handle (Mac and Windows)

Quite frequently you may want to perform the same calculation on data contained in cells of neighboring rows or columns. Instead of retyping the formula, you can simply drag the formula into adjacent cells. To do so, follow these steps:

1. Click the cell containing the formula you wish to copy.
2. Locate the fill handle, the small square in the lower right corner of the cell (see Figure A2.2).
3. Move the mouse over the fill handle to display a black plus sign (+).
4. Hold down the left mouse button and drag the fill handle over the cells you want to "fill" with the formula.

5. Click one of the "filled" cells to make sure the formula was copied correctly. If necessary, edit the formula on the formula bar. Relative cell references are the default; see Table A2.3 to make cell references absolute.

#REF! error (Mac and Windows)

When you select **Copy**, Excel copies everything in the cell—the formula, the number, and the text. When you select **Paste**, however, you may get a "#REF!" error instead of the entry you expected. This error typically occurs when the connection between the cell references and the formula is lost. To paste the value without the formula, click **Home | Clipboard ▼ | Paste ▼ | Paste Values**.

A2.4 Creating Charts

In this section, you will learn how to make three common types of graphs: XY graphs, bar graphs, and pie graphs.

XY Graphs

XY graphs are also called line graphs, but you should not confuse line graphs with Excel's "line charts." **Line charts** are a special kind of XY graph used to show trends over regular time intervals. Line charts do not space data proportionally on the x-axis. For example, intervals of 5, 20, and 50 units would be spaced equally, when in fact there should be 5 units in the first interval, 15 in the second, and 30 in the third. The takeaway message is: *When you want to make an XY graph in Excel, choose "scatter charts."*

XY graphs are used to display a relationship between two or more quantitative variables. Before you enter data in an Excel worksheet, you must determine which variable to plot on the x-axis and which one on the y-axis. By convention, the x-axis of the graph shows the independent variable, the one that was manipulated during the experiment. The y-axis of the graph shows the dependent variable, the variable that changes in response to changes in the independent variable.

Mac

Enter the data

One data set. In the first row, enter a short, informative heading for each column. Enter the values for the independent variable in column A. The independent variable is plotted on the x-axis. Enter the corresponding values for the dependent variable in column B. The dependent variable responds to the independent variable and is plotted on the y-axis.

More than one data set. If you want to plot more than one data set (line) on the same graph, follow the instructions for entering data for one data

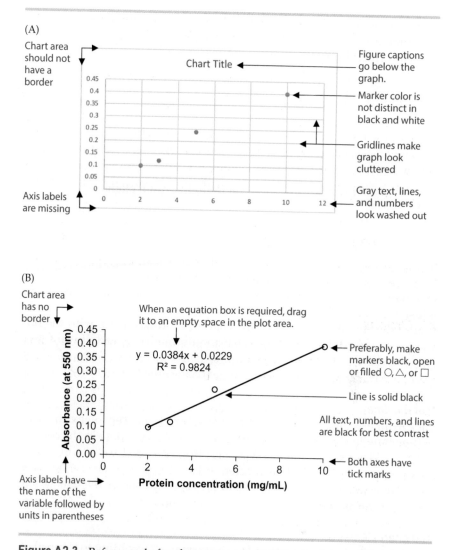

(A)

Chart area should not have a border

Chart Title

Figure captions go below the graph.

Marker color is not distinct in black and white

Gridlines make graph look cluttered

Axis labels are missing

Gray text, lines, and numbers look washed out

(B)

Chart area has no border

When an equation box is required, drag it to an empty space in the plot area.

$y = 0.0384x + 0.0229$
$R^2 = 0.9824$

Absorbance (at 550 nm)

Preferably, make markers black, open or filled ○, △, or □

Line is solid black

All text, numbers, and lines are black for best contrast

Both axes have tick marks

Axis labels have the name of the variable followed by units in parentheses

Protein concentration (mg/mL)

Figure A2.3 Before- and after-formatting of a protein standard curve. (A) The default XY graph that Excel creates has some unnecessary elements (chart border, chart title, and gridlines), and the color scheme contrasts poorly with the white background. (B) Formatting adds essential elements, eliminates clutter, and improves contrast.

set, and then enter the values for the other treatment groups in columns C, D, E, and so on. When plotting multiple data sets on the same graph, it is especially important to enter informative headings for each column, as Excel uses these headings to make a legend. Make the headings descriptive enough to allow the conditions to be distinguished unambiguously, but also keep the headings short. Using Figure A2.4 as an example, notice the difference between a concise legend entry and one that contains too much information.

FAULTY: Absorbance at 420 nm for WT, Gly

CONCISE: WT, Gly

Similarly, for Figure 5.7, it is not necessary to include the name of the dependent variable in the legend, because the reader can get this information by looking at the *y*-axis label.

FAULTY: Height in cm for light-grown plants

CONCISE: Light-grown

Select the data
Hold down the left mouse button and select the data.
One data set. Select the data (column headings are optional). When there is just one data set, there is no need for a legend.
More than one data set. Select the column headings along with the numerical values so that Excel can make an informative legend. If you select only the values, Excel will create uninformative labels like *Series 1* and *Series 2* for the legend (to edit these, see "To change the legend titles").

Plot the data
For the time being, click **Insert | X Y (Scatter) | Scatter** to plot the data without any lines. *Scatter* (as opposed to *Line*) plots the data in the correct intervals. When there are measured data, the **markers** (Excel's term for the data points) *must* be displayed. Later, in the "Choose a line" section, we will determine what kind of line (if any) is appropriate for the data based on the purpose of the graph.

Add chart elements
The graph that Excel creates needs to be formatted according to Council of Science Editors' standards (see Figures A2.3A and A2.4A). Click the graph to activate the **Chart Design** tab on the Ribbon. Click the **Add Chart Element** button all the way to the left, and make the following adjustments.

1. Axes: **Primary Horizontal** and **Primary Vertical** are checked by default.
2. Axis Titles: Click **Primary Horizontal** and enter a label for the *x*-axis. Type the variable and then the units in parentheses. Click **Primary Vertical** and enter a label for the *y*-axis. The default font color is gray. To make the font black for best contrast

(A)

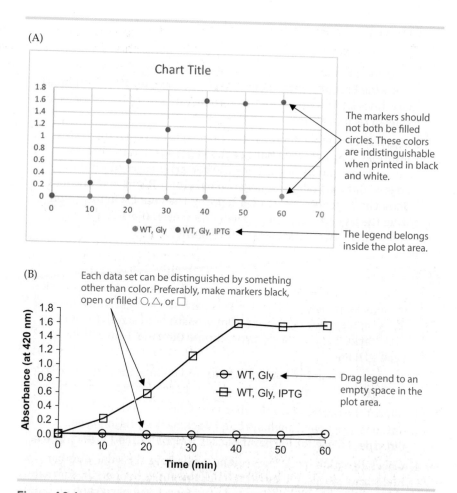

The markers should not both be filled circles. These colors are indistinguishable when printed in black and white.

The legend belongs inside the plot area.

(B)

Each data set can be distinguished by something other than color. Preferably, make markers black, open or filled ○, △, or □

Drag legend to an empty space in the plot area.

Figure A2.4 Before- and after-formatting of an XY graph with more than one line. (A) Excel's default markers are different colored circles. To make figures accessible to color blind people and those who read papers printed in black and white, use different marker shapes and line patterns to differentiate data sets. For other formatting issues, see Figure A2.3A. (B) Different marker shapes eliminate ambiguity. Black font on a white background improves contrast. The legend is an integral component of graphs with multiple data sets. For this reason, it is appropriate to place the legend inside the plot area, not outside of the axes.

on a white background, triple-click the axis label to open the Format Axis Title pane on the right. Click **Text Options | Text Fill | Color: Black**.

3. Chart Title: Change to **None** when the figure will be inserted into a scientific paper. When the figure will be used in a PowerPoint presentation, you may wish to keep **Chart Title | Above Chart**.

4. Data Labels: **None** is the default.

5. Error Bars: **None** is the default.

6. Gridlines: Remove both horizontal and vertical gridlines by unchecking **Primary Major Horizontal** and **Primary Major Vertical**. Gridlines should not be included if the purpose of the graph is to show a relationship between the variables.

7. Legend: **None** when there is only one data set. When there are multiple data sets, click **More Legend Options**. On the Format Legend pane on the right, click the **bar graph symbol** on the **Legend Options** tab, click **Legend Position | Right** and uncheck **Show the legend without overlapping the chart**. On the graph, drag the legend completely inside the axes. If there is not enough space, identify the data sets in the figure caption.

Format the axes

1. Notice that neither axis has tick marks. To insert tick marks, click the axis to display the Format Axis pane to the right. Click the **bar graph symbol** on the **Axis Options** tab, and scroll down to **Tick Marks**. Next to **Major type**, choose **Outside**. Leave **Minor type** as None.

2. Now click the **paint bucket symbol** and change the **Line** color (*not* the **Fill** color) of the axis from gray to **black** for best contrast.

3. Repeat the process for the other axis. Click the axis and **Axis Options | the bar graph symbol | Tick Marks | Major type: Outside**. Then click the **paint bucket symbol | Line | Color: Black**.

4. Check that all the numbers on the y-axis have the same number of decimal places. If they don't, click the y-axis to open the Format Axis pane to the right. Click **Axis Options | the bar graph symbol | Number**. Change the **Category** from **General** to **Number**, and change the **Decimal places** to **1** or another value that makes sense for the scale of the graph.

5. The data should fill the plot area so that there is little empty space. If necessary, change the limits of the axes by clicking the **bar graph symbol** in the **Format Axis | Axis Options** pane. Then enter the appropriate numbers in **Axis Options | Bounds | Minimum** or **Maximum**. The tick mark intervals can be adjusted under **Units | Major**.

Format the markers

Now let's change the markers from filled colored circles to symbols recommended by the Council of Science Editors for ease of recognition in black and white publications. The CSE hierarchy of symbols is open black circle, filled black circle, open black triangle, filled black triangle, open black square, filled black square (Table A2.5).

TABLE A2.5 Comparison of Excel and CSE Manual symbol hierarchy (Peterson 1999) for XY graphs	
Excel	CSE Manual
Dark blue filled circle	Black open circle
Red filled circle	Black filled circle
Gray filled circle	Black open triangle
Yellow filled circle	Black filled triangle
Light blue filled circle	Black open square
Green filled circle	Black filled square

1. Click any of the markers to display the Format Data Series pane to the right.

2. Click the **paint bucket symbol | Marker | Fill | Color: White**. This selection makes an un-filled shape. Select **Solid fill | Color: Black** to make a filled shape.

3. In the same pane, click **Border | Solid line | Color: Black**. This selection makes the outline of the symbol black for best contrast on a white background.

4. Finally, to change the size and shape of the marker itself, click the **paint bucket symbol | Marker | Marker Options | Built-in**. Select the type (circle, triangle, or square) and increase the size to at least **7**.

Choose a line

What kind of line (if any) should you use to best show the relationship between the variables? That depends on the purpose of the graph and your instructor's instructions. Reasons to add a line include

- To show a trend.
- To show a cause and effect relationship between the independent and dependent variables.

If you decide to add a line, should that line be straight or smoothed, or is a mathematical trendline the best option?

- When plotting measured data for which there is **no known mathematical relationship**, connect the points with **straight** lines. Adding **straight** lines between the markers implies that, since you didn't actually measure the in-between conditions, you are not making any inferences about them. To insert straight lines to connect the markers, click the **chart area** (the part of the graph outside the axes), and select **Chart Design | Change Chart Type | X Y (Scatter) | Scatter with Straight Lines and Markers**.
- If the intermediate values can be interpolated with confidence, it may be appropriate to use a **smoothed** line. To insert smooth lines

to connect the markers, click the **chart area**, and select **Chart Design | Change Chart Type | X Y (Scatter) | Scatter with Smooth Lines and Markers**.

- Finally, when there is an **expected mathematical relationship** between the variables, insert a **trendline**. For example, from Beer's Law we expect absorbance to be proportional to concentration. Thus a **linear** trendline should be used for these data. On the other hand, enzyme kinetics has established that enzymatic activity increases logarithmically as a function of substrate concentration. Thus a **logarithmic** trendline should be inserted on these data points.

To insert a trendline, click a marker and under **Chart Design | Add Chart Element**, select **Trendline | More Trendline Options**.

1. On the Format Trendline pane to the right, click **the bar graph symbol**. You'll see that **Linear** is the default, but other options are available.

2. If you are using this graph to predict one variable from the other variable, which is usually measured, display the equation of the line. In addition, to show how well (or not) the equation fits the experimental data, display the R-squared value. To do so, check the **Display Equation on chart** and **Display R-squared value on chart** boxes at the bottom of the pane.

3. On the graph itself, drag the text box so that it doesn't overlap the trendline. Highlight the text and change the font color from gray to black.

Format the line

Solid black lines are preferred for best contrast on a white background.
Scatter with straight/smooth lines and markers. To change Excel's default colored lines to solid black lines, click the line to display the Format Data Series pane on the right. Click the **paint bucket symbol | Line** and change the color to **black**.
Trendlines. Click the trendline to display the Format Trendline pane. Click the **paint bucket symbol** and change the color to **black**. Under **Dash type**, choose the first option, the **solid line**. When there are multiple lines on the same graph, different dash types may be used to help distinguish the data sets.

Remove chart border

Figures in scientific papers do not have a border. To remove the border around the graph, click the chart area (*not* the **plot area**, which is the area inside the axes). In the Format Chart Area pane to the right, click **Chart Options | the paint bucket symbol**, and under **Border**, click **No line**.

Save as a chart template

To save the XY graph format as a chart template, see Section A2.5.

Import graph into Word document

Compare the before- and after-formatting graphs in Figures A2.3 and A2.4. The gridlines in the before-graph make the graph look cluttered and obscure the trend shown by the data points. The colored markers (which look gray when the graph is printed in a black and white publication) do not stand out against a white background. The crisp format of the after-graph focuses attention on the results.

To copy and paste the graph you made in Excel into your Word document, follow these steps.

1. Click the chart area (not the plot area).

2. Click **Home | Copy** or ⌘+**C**.

3. In your Word document, position the cursor below the paragraph in which you first describe the graph (typically in the Results section). Click **Home | Paste** or ⌘+**V**.

4. Press **Return** after inserting the figure, and then type a figure caption *below* the figure.

5. If you would like to resize the graph, but maintain the 3×5 aspect ratio, hold down the **Shift** key when dragging the corner.

Windows

Enter the data

One data set. In the first row, enter a short, informative heading for each column. Enter the values for the independent variable in column A. The independent variable is plotted on the x-axis. Enter the corresponding values for the dependent variable in column B. The dependent variable responds to the independent variable and is plotted on the y-axis.

More than one data set. If you want to plot more than one data set (line) on the same graph, follow the instructions for entering data for one data set, and then enter the values for the other treatment groups in columns C, D, E, and so on. When plotting multiple data sets on the same graph, it is especially important to enter informative headings for each column, as Excel uses these headings to make a legend. Make the headings descriptive enough to allow the conditions to be distinguished unambiguously, but also keep the headings short. Using Figure A2.4 as an example, notice the difference between a concise title for the legend and one that contains too much information.

FAULTY: Absorbance at 420 nm for WT, Gly

CONCISE: WT, Gly

Similarly, for Figure 5.7, it is not necessary to include the name of the dependent variable in the legend, because the reader can get this information by looking at the y-axis label.

FAULTY: Height in cm for light-grown plants

CONCISE: Light-grown

Select the data

Hold down the left mouse button and select the data.

One data set. Select the data (column headings are optional). When there is just one data set, there is no need for a legend.

More than one data set. Select the column headings along with the numerical values so that Excel can make an informative legend. If you select only the values, Excel will create uninformative labels like *Series 1* and *Series 2* for the legend (to edit these, see "To change the legend titles").

Plot the data

For the time being, click **Insert | Insert Scatter (X, Y) or Bubble Chart ▼ | Scatter** to plot the data without any lines. *Scatter* (as opposed to *Line*) plots the data in the correct intervals. When there are measured data, the **markers** (Excel's term for the data points) *must* be displayed. Later, in the "Choose a line" section, we will determine what kind of line (if any) is appropriate for the data based on the purpose of the graph.

Add chart elements

The graph that Excel creates needs to be formatted according to Council of Science Editors' standards (compare the before (A) and after (B) graphs in Figures A2.3 and A2.4). Click the graph to activate the 3 buttons at the top right of the graph. Click the **+ Chart Elements** button to display a checklist of formatting elements. The same list can be accessed by clicking **Chart Tools Design | Add Chart Element**.

1. Axes: **Primary Horizontal** and **Primary Vertical** are checked by default.
2. Axis Titles: Click **Axis Titles**. Enter a label for the y-axis. Type the variable and then the units in parentheses. In the same way, enter a label for the x-axis. Triple-click each label and change the font color from gray to **black** on the mini toolbar.
3. Chart Title: Change to **None** when the figure will be inserted into a scientific paper. When the figure will be used in a PowerPoint presentation, you may wish to keep **Chart Title | Above Chart**.
4. Data Labels: **None** is the default.
5. Error Bars: **None** is the default.

6. Gridlines: Uncheck **Gridlines**. Gridlines should not be included if the purpose of the graph is to show a relationship between the variables.

7. Legend: **None** when there is only one data set. When there are multiple data sets, click **More Options**. Select **Legend Position | Right** on the Format Legend pane to the right, and uncheck **Show the legend without overlapping the chart**. On the graph, drag the legend completely inside the axes. If there is not enough space, identify the data sets in the figure caption.

Format the axes

1. Notice that neither axis has tick marks. To insert tick marks, click the axis to display the Format Axis pane to the right. Click the **bar graph symbol** on the **Axis Options** tab and scroll down to **Tick Marks**. Next to **Major type**, choose **Outside**. Leave **Minor type** as **None**.

2. Now click the **paint bucket symbol** and change the **Line** color (*not* the **Fill** color) of the axis from gray to **black**.

3. Repeat the process for the other axis. Click the axis, and click **Axis Options | the bar graph symbol | Tick Marks | Major type: Outside**. Then click the **paint bucket symbol | Line | Color: Black**.

4. Check that all the numbers on the y-axis have the same number of decimal places. If they don't, click the y-axis to open the Format Axis pane. Click **Axis Options | the bar graph symbol | Number**. Change the **Category** from **General** to **Number**, and change the **Decimal places** to **1** or another value that makes sense for the scale of the graph.

5. The data should fill the plot area so that there is little empty space. If necessary, change the limits of the axes by clicking the **bar graph symbol** in the **Format Axis | Axis Options** pane. Then enter the appropriate numbers in **Axis Options | Bounds | Minimum** or **Maximum**. The tick mark intervals can be adjusted under **Units | Major**.

Format the markers

Now let's change the markers from filled colored circles to symbols recommended by the Council of Science Editors for ease of recognition in black and white publications. The CSE hierarchy of symbols is unfilled (open) black circle, filled black circle, unfilled black triangle, filled black triangle, unfilled black square, filled black square (Table A2.5).

1. Click any of the markers to display the Format Data Series pane to the right.

2. Click the **paint bucket symbol | Marker | Fill | Color: White**. This selection makes an unfilled shape. Select **Solid fill | Color: Black** to make a filled shape.

3. In the same pane, click **Border | Solid line | Color: Black**. This selection makes the outline of the symbol black for best contrast on a white background.

4. Finally, to change the size and shape of the marker itself, click the **paint bucket symbol | Marker | Marker Options | Built-in**. Select the type (circle, triangle, or square) and increase the size to at least **7**.

Choose a line

What kind of line or curve should you use to best show the relationship between the variables? That depends on the purpose of the graph and your instructor's instructions. Reasons to add a line include

- To show a trend.
- To show a cause and effect relationship between the independent and dependent variables.

If you decide to add a line, should that line be straight or smoothed, or is a mathematical trendline the best option?

- When plotting measured data for which there is **no known mathematical relationship**, connect the points with **straight** lines. Adding **straight** lines between the markers implies that, since you didn't actually measure the in-between conditions, you are not making any inferences about them. To insert straight lines to connect the markers, click the **chart area** (the part of the graph outside the axes), and select **Chart Tools Design | Change Chart Type | X Y (Scatter) | Scatter with Straight Lines and Markers**.

- If the intermediate values can be interpolated with confidence, it may be appropriate to use a **smoothed** line. To insert smooth lines to connect the markers, click the **chart area** (the part of the graph outside the axes), and select **Chart Tools Design | Change Chart Type | X Y (Scatter) | Scatter with Smooth Lines and Markers**.

- Finally, when there is an **expected mathematical relationship** between the variables, insert a **trendline**. For example, from Beer's Law we expect absorbance to be proportional to concentration. Thus a **linear** trendline should be used for these data. On the other hand, enzyme kinetics has established that enzymatic activity increases logarithmically as a function of substrate concentration. Thus a **logarithmic** trendline should be inserted on these data points.

To insert a trendline, click a marker and under **Chart Tools Design | Add Chart Element**, select **Trendline | More Trendline Options**.

1. On the Format Trendline pane to the right, click the **bar graph symbol**. You'll see that **Linear** is the default, but other options are available.

2. If you are using this graph to predict one variable from the other variable, which is usually measured, display the equation of the line. In addition, to show how well (or not) the equation fits the experimental data, display the R-squared value. To do so, check the **Display Equation on chart** and **Display R-squared value on chart** boxes at the bottom of the pane.

3. On the graph itself, drag the text box so that it doesn't overlap the trendline. Highlight the text and change the font color from gray to black.

Format the line
Solid black lines are preferred for best contrast on a white background. **Scatter with straight/smooth lines and markers.** To change Excel's default colored lines to solid black lines, click the line to display the Format Data Series pane on the right. Click the **paint bucket symbol | Line** and change the color to **black**.

Trendlines.
Click the trendline to display the Format Trendline pane. Click the **paint bucket symbol** and change the color to **black**. Under **Dash type**, choose the first option, the **solid line**. When there are multiple lines on the same graph, different dash types may be used to help distinguish the data sets.

Remove chart border
Figures in scientific papers do not have a border. To remove the border around the graph, click the chart area (*not* the **plot area**, which is the area inside the axes). In the Format Chart Area pane to the right, click **Chart Options | the paint bucket symbol**, and under **Border**, click **No line**.

Save as a chart template
To save the XY graph format as a chart template, see Section A2.5.

Import graph into Word document
Compare the before- and after-formatting graphs in Figures A2.3 and A2.4. The gridlines in the before-graph make the graph look cluttered and obscure the trend shown by the data points. The colored markers (which look

gray when the graph is printed in a black and white publication) do not stand out against a white background. The crisp format of the after-graph focuses attention on the results.

To copy and paste the graph you made in Excel into your Word document, follow these steps.

1. Click the chart area (not the plot area).

2. Click **Home | Copy** or **Ctrl+C**.

3. In your Word document, position the cursor below the paragraph in which you first describe the graph (typically in the Results section). Click **Home | Paste** or **Ctrl+V**.

4. Press **Enter** after inserting the figure, and then type a figure caption *below* the figure.

5. If you would like to resize the graph, but maintain the 3x5 aspect ratio, hold down the **Shift** key when dragging the corner.

XY graphs with multiple lines

Plotting multiple lines on one graph is often the most efficient way to compare the results from several different treatments. How many lines should you put on one set of axes? The CSE Manual recommends no more than eight, but use common sense. You should be able to follow each line individually, and the graph should not look cluttered.

Follow the instructions for entering and selecting the data in the "XY Graphs" section. When formatting the graph, do not use Excel's defaults for the markers. Instead, replace the colored circles with open or filled ○, △, or □, and include a legend for identification (see Figure A2.4).

Bar graphs

Bar graphs are used to compare individual data sets when one of the variables is categorical (not quantitative). By convention, the feature that all the columns have in common (the variable that was measured) lies on the axis parallel to the columns. Excel distinguishes between two basic types of bar graphs: column charts and bar charts. **Column charts** are bar graphs with vertical bars. **Bar charts** are bar graphs with horizontal bars. Bar charts are more practical than column charts when the category labels are long (compare Figures A2.5A and A2.5B).

Enter the labels for the categories in Column A and the quantitative data in Column B. The categories should be sequential on the graph, with the control treatment column farthest left. When deciding what order to enter the categories in the worksheet, remember that the *lowest* row number contains the category label for the *leftmost* vertical bar or the *lowest* horizontal bar. If there is no particular order to the categories, arranging the bars from shortest to longest (or vice versa) makes the results easier to comprehend.

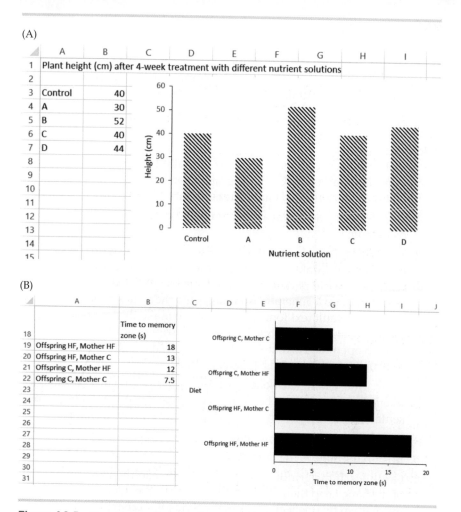

(A)

	A	B	C	D	E	F	G	H	I
1	Plant height (cm) after 4-week treatment with different nutrient solutions								
2									
3	Control	40							
4	A	30							
5	B	52							
6	C	40							
7	D	44							
8									
9									
10									
11									
12									
13									
14									
15									

(B)

	A	B	C	D	E	F	G	H	I	J
18		Time to memory zone (s)								
19	Offspring HF, Mother HF	18								
20	Offspring HF, Mother C	13								
21	Offspring C, Mother HF	12								
22	Offspring C, Mother C	7.5								
23										
24										
25										
26										
27										
28										
29										
30										
31										

Figure A2.5 Two types of bar graphs. For both types, the labels for the categorical variable are entered in Column A; the data for the response variable are entered in Column B. (A) Column charts have vertical columns. The data for the control are entered in the lowest numbered row to position the corresponding column farthest left on the graph. This column chart has no baseline on the categorical axis. The requirements of the publication (or your instructor) determine whether the line should be present or not. (B) Bar charts have horizontal columns. The category label for the lowest bar (and its data) is entered in the lowest numbered row. Bar charts accommodate long category labels. This bar chart has a baseline on the categorical axis. Gridlines may be included to facilitate reading the values of the bars.

Mac

Enter and select the data

1. In the first row, enter a short, informative heading for each column. Enter the categories in column A. Enter the corresponding values for the quantitative variable in column B.

2. Hold down the left mouse button and select the data (including the column headings). To make a column chart with vertical bars, click **Insert | Column | Clustered Column**. To make a bar chart with horizontal bars, click **Insert | Bar | Clustered Bar**.

Add chart elements

The graph that Excel creates needs to be formatted according to Council of Science Editors' standards. Click the graph to activate the **Chart Design** tab on the Ribbon. Click the **Add Chart Element** button all the way to the left.

1. Axes: **Primary Horizontal** and **Primary Vertical** are checked by default.

2. Axis Titles: The label for the response axis (the one with the quantitative variable) should have the name of the variable with the units in parentheses. The label for the categorical axis should be a general term that covers the specific categories. The default text font color is gray. To make the font black for best contrast on a white background, triple-click the axis label to open the Format Axis Title pane on the right. Click **Text Options | Text Fill | Color: Black**.

3. Chart Title: Change to **None** when the figure will be inserted into a scientific paper. When the figure will be used in a PowerPoint presentation, you may wish to keep **Chart Title | Above Chart**.

4. Data Labels: **None** is the default.

5. Error Bars: **None** is the default.

6. Gridlines: If the gridlines make it easier to read the value of the bars, keep them. If they compete with the data, remove them by unchecking **Primary Major Horizontal** and **Primary Major Vertical**.

7. Legend: **None** when there is only one data set. When there are multiple data sets (as in clustered column charts), click **More Legend Options**. Select **Legend Position | Right** and uncheck **Show the legend without overlapping the chart**. On the graph, drag the legend completely inside the axes. If there is not enough space, identify the data sets in the figure caption.

Format the axes

1. Notice that the axis with the quantitative (response) variable does not have tick marks. To insert tick marks, click the axis to display the Format Axis pane to the right. Click the **bar graph symbol**, and scroll down to **Tick Marks**. Next to **Major type**, choose **Outside**. Leave **Minor type** as **None**.

2. Now click the **paint bucket symbol** and change the **Line** color (*not* the **Fill** color) of the axis from gray to **black** for best contrast.

3. According to the CSE Manual, the axis with the categorical variable may or may not be visible, but all the columns must be aligned as if there were a baseline. To remove the line, click the axis to open the Format Axis pane. Then click **Axis Options | the paint bucket symbol | Line | No line**. Alternatively, make the line black by clicking **Line | Color: Black.** To make the text font black, click **Text Options | Text Fill | Color: Black.**

Adjust bar width

All of the bars in the graph should be the same width, and the bars should always be wider than the space between them. To adjust the width of the bars, click any one of them to open the Format Data Series pane. Click **the bar graph symbol,** and drag the slider for **Gap width** to the left to decrease the gap and simultaneously increase the width of the bars.

Change bar color

Still in the Format Data Series pane, click **the paint bucket symbol | Fill | Solid fill | Color: Black** to make the bars black against a white background for best contrast in black-and-white publications. To make white bars with a black border, click **the paint bucket symbol | Fill | No fill** and then **Border | Solid line | Color: Black.** Alternatively, for good contrast with less ink, click **the paint bucket symbol | Fill | Pattern Fill** and choose one of the heavier line patterns. Change the foreground color to black.

Remove chart border

Figures in scientific papers do not have a border. To remove the border around the graph, click the chart area (*not* the **plot area,** which is the area inside the axes). In the Format Chart Area pane to the right, click **Chart Options | the paint bucket symbol,** and under **Border,** click **No line.**

Save as a chart template

To save the bar graph format as a chart template, see Section A2.5.

Import graph into Word document

To copy and paste the graph you made in Excel into your Word document, follow these steps.

1. Click the chart area (not the plot area).

2. Click **Home | Copy** or ⌘+**C.**

3. In your Word document, position the cursor below the paragraph in which you first describe the graph (typically in the Results section). Click **Home | Paste** or ⌘+V.

4. Press **Return** after inserting the figure, and then type a figure caption *below* the figure.

5. If you would like to resize the graph, but maintain the 3x5 aspect ratio, hold down the **Shift** key when dragging the corner.

Windows

Enter and select the data

1. In the first row, enter a short, informative heading for each column. Enter the categories in column A. Enter the corresponding values for the quantitative variable in column B.

2. Hold down the left mouse button and select the data (including the column headings). To make a column chart with vertical bars, click **Insert | Charts | Insert Column Chart | 2-D Column.** To make a bar chart with horizontal bars, click **Insert | Charts | Insert Bar Chart | 2-D Bar**.

Add chart elements

The graph that Excel creates needs to be formatted according to Council of Science Editors' standards. Click the graph to activate the 3 buttons at the top right of the graph. Click the **+ Chart Elements** button to display a checklist of formatting elements.

1. Axes: **Primary Horizontal** and **Primary Vertical** are checked by default.

2. Axis Titles: The label for the response axis (the one with the quantitative variable) should have the name of the variable with the units in parentheses. The label for the categorical axis should be a general term that covers the specific categories. The default text font color is gray. Triple-click each label and change the font color from gray to **black** on the mini toolbar.

3. Chart Title: Change to **None** when the figure will be inserted into a scientific paper. When the figure will be used in a PowerPoint presentation, you may wish to keep **Chart Title | Above Chart**.

4. Data Labels: **None** is the default.

5. Error Bars: **None** is the default.

6. Gridlines: If the gridlines make it easier to read the value of the bars, keep them. If they compete with the data, remove them by unchecking **Gridlines**.

7. Legend: **None** when there is only one data set. When there are multiple data sets (as in clustered column charts), click **More Options**. In the Format Legend pane, click the **bar graph symbol**. Select **Legend Position | Right** and uncheck **Show the legend without overlapping the chart**. On the graph, drag the legend completely inside the axes. If there is not enough space, identify the data sets in the figure caption.

Format the axes

1. Notice that the axis with the quantitative (response) variable does not have tick marks. To insert tick marks, click the axis to display the Format Axis pane to the right. Click the **bar graph symbol**, and scroll down to **Tick Marks**. Next to **Major type**, choose **Outside**. Leave **Minor type** as **None**.

2. Now click the **paint bucket symbol** and change the **Line** color (*not* the **Fill** color) of the axis from gray to **black** for best contrast.

3. According to the CSE Manual, the axis with the categorical variable may or may not be visible, but all the columns must be aligned as if there were a baseline. To remove the line, click the axis to open the Format Axis pane. Then click **Axis Options | the paint bucket symbol | Line | No line**. Alternatively, make the line black by clicking **Line | Color: Black**. To make the text font black, click **Text Options | Text Fill | Color: Black**.

Adjust bar width

All of the bars in the graph should be the same width, and the bars should always be wider than the space between them. To adjust the width of the bars, click any one of them to open the Format Data Series pane. Click **the bar graph symbol**, and drag the slider for **Gap Width** to the left to decrease the gap and simultaneously increase the width of the bars.

Change bar color

Still in the Format Data Series pane, click **the paint bucket symbol | Fill | Solid fill | Color: Black** to make the bars black against a white background for best contrast in black-and-white publications. To make white bars with a black border, click the **paint bucket symbol | Fill | No fill** and then **Border | Solid line | Color: Black**. Alternatively, for good contrast with less ink, click the **paint bucket symbol | Fill | Pattern Fill** and choose one of the heavier line patterns. Change the foreground color to black.

Remove chart border

Figures in scientific papers do not have a border. To remove the border around the graph, click the chart area (*not* the **plot area**, which is the area inside the axes). In the Format Chart Area pane to the right, click **Chart Options | the paint bucket symbol**, and under **Border**, click **No line**.

Save as a chart template
To save the bar graph format as a chart template, see Section A2.5.

Import graph into Word document
To copy and paste the graph you made in Excel into your Word document, follow these steps.

1. Click the chart area (not the plot area).
2. Click **Home | Copy** or **Ctrl+C**.
3. In your Word document, position the cursor below the paragraph in which you first describe the graph (typically in the Results section). Click **Home | Paste** or **Ctrl+V**.
4. Press **Enter** after inserting the figure, and then type a figure caption *below* the figure.
5. If you would like to resize the graph, but maintain the 3x5 aspect ratio, hold down the **Shift** key when dragging the corner.

Clustered column charts

Clustered columns may represent the results of different treatments after the same period of time or the results of the same treatments after different periods of time (Figure A2.6). Each column in the cluster must be easy to distinguish from its neighbor. Colorful columns that look good on your computer screen may turn out to be the same shade of gray when the graph is printed on a black-and-white printer. If the person who evaluates your work receives a black-and-white copy of your paper, be sure to proofread the hardcopy and check that it's clear which treatments the columns represent.

To make a clustered column chart, follow the instructions for plotting bar graphs, with the following differences:

- The response (quantitative) values for the bars in each cluster are entered in Column B, C, and so on.
- When entering column headings for the data, do not enter a heading for Column A, otherwise Excel will not format the graph correctly.
- To adjust the width and spacing of the bars, click one of them to open the Format Data Series pane. Click **the bar graph symbol**, and drag the slider for **Gap Width** to the left to increase the width of the columns. In the same pane, adjust **Series Overlap** to **0%** to eliminate the space between the columns in each cluster.

Leave column heading
for category blank

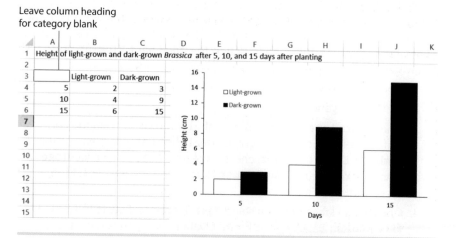

Figure A2.6 Final form of a clustered column chart after formatting in Excel. Additional changes can be made in a Word document after copying and pasting the graph from Excel.

Pie graphs

Pie graphs are commonly used to show financial data, but they are seldom used in biology research papers. The CSE Manual recommends a table rather than a pie graph for showing percentage of constituents out of the whole. If you would like to make a pie graph, however, follow these instructions.

Mac

Enter and select the data

1. In the first row, enter a short, informative heading for each column. These column headings are for your information only; they are not used to make the pie chart. Enter the categories in column A. Enter the corresponding values for the quantitative variable in column B. Sort the data (see Section A2.2) according to percentage, with the highest percentage in the first row. This arrangement will result in the largest segment of the pie beginning at 12 o'clock, with the segments decreasing in size clockwise.

2. Hold down the left mouse button and select the data (the result is the same with or without the column headings). Click **Insert | Pie | 2-D Pie**.

Add chart elements

The graph that Excel creates needs to be formatted according to Council of Science Editors' standards (Figure A2.7). Click the graph to activate the **Chart Design** tab on the Ribbon. Click the **Add Chart Element** button all the way to the left.

1. Uncheck **Chart Title**.

2. Uncheck **Legend**.

3. Check **Data Labels | More Data Label Options** to open the Format Data Labels pane on the right. Under **Label Options | bar graph symbol | Label Contains**, check **Category Name** and **Percentage**. Uncheck **Value** and **Show Leader Lines**. Under **Label Position**, check **Outside End**.

4. Click a data label and then select **Text Options | Text Fill | Solid fill | Color: Black** on the Format Data Labels pane on the right.

Remove chart border

Figures in scientific papers do not have a border. To remove the border around the graph, click the chart area. In the Format Chart Area pane to the right, click **Chart Options | the paint bucket symbol**, and under **Border**, click **No line**.

Save as a chart template

To save the pie chart format as a chart template, see Section A2.5.

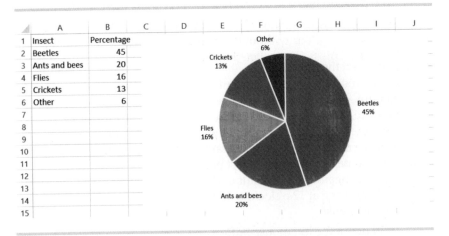

Figure A2.7 Final form of a pie chart after formatting in Excel. Additional changes can be made in a Word document after copying and pasting the graph from Excel.

Import graph into Word document

To copy and paste the graph you made in Excel into your Word document, follow these steps.

1. Click the chart area.
2. Click **Home | Copy** or ⌘+C.
3. In your Word document, position the cursor below the paragraph in which you first describe the graph (typically in the Results section). Click **Home | Paste** or ⌘+V.
4. Press **Return** after inserting the figure, and then type a figure caption *below* the figure.
5. If you would like to resize the graph, but maintain the 3x5 aspect ratio, hold down the **Shift** key when dragging the corner.

Windows

Enter and select the data

1. In the first row, enter a short, informative heading for each column. These column headings are for your information only; they are not used to make the pie chart. Enter the categories in column A. Enter the corresponding values for the quantitative variable in column B. Sort the data (see Section A2.2) according to percentage, with the highest percentage in the first row. This arrangement will result in the largest segment of the pie beginning at 12 o'clock, with the segments decreasing in size clockwise.
2. Hold down the left mouse button and select the data (the result is the same with or without the column headings). Click **Insert | Charts | Insert Pie or Doughnut Chart | 2-D Pie**.

Add chart elements

The graph that Excel creates needs to be formatted according to Council of Science Editors' standards. Click the graph to activate the 3 buttons at the top right of the graph. Click the **+ Chart Elements** button to display a checklist of formatting elements.

1. Uncheck **Chart Title**.
2. Uncheck **Legend**.
3. Check **Data Labels | Data Callout**.
4. To remove the callout borders, first click a data callout to select all of the callouts. Then, in the Format Data Labels pane to the right, click **Label Options | the paint bucket symbol**, and under **Border**, click **No line**.

5. To change the lettering from gray to black in the callouts, click a data callout as before. Then select **Font | Font color: Black** from the **Home** tab on the Ribbon.

Remove chart border

Figures in scientific papers do not have a border. To remove the border around the graph, click the chart area. In the Format Chart Area pane to the right, click **Chart Options | the paint bucket symbol**, and under **Border**, click **No line**.

Save as a chart template

To save the pie chart format as a chart template, see Section A2.5.

Import graph into Word document

To copy and paste the graph you made in Excel into your Word document, follow these steps.

1. Click the chart area.
2. Click **Home | Copy** or **Ctrl+C**.
3. In your Word document, position the cursor below the paragraph in which you first describe the graph (typically in the Results section). Click **Home | Paste** or **Ctrl+V**.
4. Press **Enter** after inserting the figure, and then type a figure caption below the figure.
5. If you would like to resize the graph, but maintain the 3x5 aspect ratio, hold down the **Shift** key when dragging the corner.

A2.5 Saving and Applying Chart Templates

It takes a fair amount of time to format graphs in Excel because there are so many options. Fortunately, you can save the format as a **chart template** to save yourself time the next time you have to make the same type of chart.

Mac

1. Click the chart and then **Chart Design | Change Chart Type | Save As Template**. Give the template a specific name, for example, **Scatter with 1 linear trendline**. Do not change the default location, which is the Chart Templates folder.
2. To apply this format to new data, select the data and on the top menu bar (not the Ribbon), click **Insert | Chart | Templates**. Select the desired saved template.

3. For some strange reason, when scatter chart templates are applied to new data, Excel creates a chart with incorrect formatting. To correct the problem, click the chart, and then click **Chart Design | Switch Row/Column** twice.

4. Finally, click each Axis Title text box and type the desired title. If necessary, adjust the length of the axes to eliminate empty space as described in "Format the axes."

Windows

1. Right-click the chart and select **Save As Template** from the drop-down menu. Give the template a specific name, for example, **Scatter with 1 linear trendline**. Do not change the default **Save as type** (Chart Template Files).

2. To apply this format to new data, select the data and click **Insert | Charts diagonal arrow | All Charts | Templates**. Double-click the saved template that you want to apply.

3. For some strange reason, when scatter chart templates are applied to new data, Excel creates a chart with incorrect formatting. To correct the problem, click the chart, and then click **Chart Tools Design | Switch Row/Column** twice.

4. Finally, click each Axis Title text box and type the desired label. If necessary, adjust the length of the axes to eliminate empty space as described in "Format the axes."

A2.6 Editing the Graph After it Has Been Formatted (Mac and Windows)

If you have gone to all this trouble to format a graph, and realize after the fact that you need to add another data point, the last thing you want to do is to start from scratch. Fortunately, Excel makes it possible to edit data after the graph has been made.

To incorporate additional data points in the same series

Insert a new row in the worksheet for the new data points. To do this, click a cell in the row below where the new row is to be inserted. Then click **Home | Cells | Insert ▼ | Insert Sheet Rows**. Enter the value for the x-axis in Column A and that for the y-axis in Column B. The graph is automatically updated for these values.

To incorporate additional lines on the same graph

When adding data sets to an existing graph, the x-axis values remain the same, but additional y-axis values that represent data for a different treatment or condition have to be added. The y-axis values for the existing line are already in Column B. Type the y-axis values for the second line in Column C. For each additional line, enter data in the next column to the right.

1. To add the new data to the existing graph, right-click anywhere in the chart area or the plot area and click **Select Data** (Windows) or **Chart Design | Select Data** (Mac).

   Mac. In the **Select Data Source** dialog box, click the **+** button below the **Legend Entries (Series)** list box.

   Windows. In the **Select Data Source** dialog box, click the **Add** button in the **Legend Entries (Series)** list box.

2. Enter a descriptive title for each data set, which will be used in the legend.

   Mac. To the right of the **Legend Entries (Series)** list box, click the box next to **Name**. Click the cell reference containing the column heading, or enter a short title for the legend.

   Windows. In the **Edit Series** dialog box, under **Series name**, click the cell reference containing the column heading, or enter a short title for the legend.

3. Enter the range of values for the x-axis.

   Mac. Click the box next to **X values**. Then click the icon with the red arrow pointing to a cell in a worksheet. Select the values in Column A. Click the icon (now with a down-arrow) again.

   Windows. Under **Series X values**, click the icon with the black up-arrow. Select the values in Column A. Click the icon (now a down-arrow) again.

4. Enter the range of values for the y-axis for the data to be added.

   Mac. Click the box next to **Y values**. Then click the icon with the red arrow pointing to a cell in a worksheet. Select the values in Column C. Click the icon (now with a down-arrow) again.

   Windows. Under **Series Y values**, click the icon with the black up-arrow. Select the values in Column C. Click the icon (now a down-arrow) again.

5. Repeat Steps 2–4 for each data set. Then Click **OK** to close the **Select Data Source** dialog box.

6. Change the new symbol(s) and line(s) as described under "Format Markers" and "Format the Line."

To change the legend titles

The easiest way to make descriptive legend titles is to make descriptive column headings for your data. Then select the headings along with the data when you make the graph. However, if you need to change the titles after your graph has been finalized, follow these instructions.

1. Click a marker of the data series whose title you want to change. In the Mac OS, click **Chart Design | Select Data**. In Windows, right-click the chart area and click **Select Data** from the drop-down menu.

2. In the Select Data Source dialog box of the Mac OS, select the content next to **Name**, and type a new name for the legend entry. Click **OK** to exit. In Windows, select the data set to be edited and then click **Legend Entries (Series) | Edit**. Under Series name, click the icon with the black up-arrow. Select the cell that contains the column heading or type a name. Click the icon again. Click **OK** twice to exit.

Go to **macmillanlearning.com/knisely6e**
and select "Student Site" to access
samples, template files, and tutorial videos

Preparing Oral Presentations with Microsoft PowerPoint

A3.1 Introduction

Microsoft PowerPoint allows you to create a slide show containing text, graphics, and audiovisual media. Besides being used as a visual aid during the presentation, the **slide deck** (the slides and any notes that make up a presentation) may be posted online as a study aid or printed out for taking notes. Slides may also be reused in other presentations and shared for documentation purposes, particularly in the corporate environment. Slide shows can also be recorded with audio synced to each slide.

The first section of this appendix emphasizes the importance of planning your presentation before you even open PowerPoint. The planning stage is where you think about what your audience already knows and how to present new information in an understandable and engaging manner. Ideally, you want to give a presentation that is worth your audience members' time.

The second part of this appendix shows you how to construct slides that are audience-friendly. Audience-friendly slides are, first of all, easy on the eyes, and, secondly, focused on a single idea. "Easy on the eyes" means not just esthetically pleasing, but legible. Text must be large enough to be read from the back of the presentation room. Images must be sharp. Both text and images must stand out on the background. Slides that focus audience attention on a single idea contain a minimum amount of text that, whenever possible, is supported with a visual aid. Animations can be used to emphasize one element of the slide at a time. Speaker prompts that potentially make slides text-heavy are entered in the Notes pane, which is seen by the speaker in Presenter View, but not by the audience.

The third part of this appendix is about delivering PowerPoint presentations. For presentations to be successful, speakers must first practice reading and refining their script until they have practically memorized it. **Presenter View** mode is useful for rehearsal as well as for keeping the speaker on track during the actual presentation. In this mode, the current slide seen by the audience is displayed on the screen along with prompts in the Notes pane, which are seen only by the speaker. PowerPoint also provides tools for navigating among slides and writing on the slides. Finally, slide decks can be shared electronically or as handouts. For all other PowerPoint questions, please visit the Microsoft Office support website.

In this book, the nomenclature for the command sequence is as follows:

> **Ribbon tab | Group** (*Windows only*) **| Command button |**
> **Additional Commands** (*if available*).

For example, to insert a slide with title and content text boxes in Windows, **Home | Slides | New Slide ▼ | Title and Content** means "Click the **Home** tab and in the **Slides** group, click the down arrow (▼) on the **New Slide** button to select the **Title and Content** option. In the Mac OS, the corresponding command sequence is **Home | New Slide ▼ | Title and Content**.

Save and back up computer files

Good file organization, as described in Section A1.2, is just as important for PowerPoint presentations as Word documents. Review this section to develop good habits for naming, saving, and backing up computer files.

A3.2 Planning Your Presentation

Before you open PowerPoint, make a concept map of your presentation (see Section 3.5). Think about what your audience already knows and how to lead them to what you want them to know. Very roughly sketch or outline the topics and decide on the order in which you want to present them. It doesn't matter if you use notebook paper, a whiteboard, or an electronic device—the important thing is to give your ideas some structure. The concept map doesn't have to be perfect. It is very common to change things around in the course of preparing a presentation.

As much as possible, make or find visuals that support the most important points of your talk. In biology talks, it is common to see photos, graphs, gel images, phylogenetic trees, and other visuals. Make sure that these figures are large, legible, and sharp when projected onto a large screen. As you prepare the visuals, write down what you want to say about them.

After preparing the individual components of your presentation, you are finally ready to open PowerPoint, choose an appropriate slide design (theme), and insert content on the slides.

A3.3 Designing Slides

Selecting a theme

When you open PowerPoint, you have a number of options. You can open an existing presentation, select a blank presentation or a colorful template or themed presentation from the gallery, or search for templates or themes online.

Mac **File | New Presentation** opens a blank presentation. **File | New from Template** opens the gallery of themes, which includes **Blank Presentation**.

Windows **File | New** opens the gallery of themes, which includes **Blank Presentation**.

Themes (designs) with a light background make the room much brighter during the actual presentation, which has the dual advantage of keeping the audience awake and allowing you to see your listeners' faces. Many professional speakers, however, prefer white text on a dark background because the text appears larger. Whatever your preference, choose a theme that reflects your style, is appropriate for the topic, and complements the content of the individual slides.

From the perspective of the audience, two non-negotiable criteria for slide design are **contrast** and **font size**. For content to be legible from a distance in a large room, the text must stand out from the background, and the font size has to be at least 24 pt. To determine if a theme meets these criteria:

Mac

1. Click inside one of the text boxes in the Slide pane (see Figure A3.1B). Click a theme on the **Design** tab. The down and right arrows display more themes. After clicking a theme, choose a color and background style from the options on the right. Keep in mind that black on a light background or white on a dark background provides the best contrast.

2. To check the default font sizes for this theme variant, click **View | Slide Master** and then scroll to the very first slide thumbnail in the navigation pane on the left. When you edit this slide, the changes will be applied to all of the slides in the presentation, not just to certain slide layouts. Click each style (title, first level text, second level text, and so on) to view the font style and size on the **Home** tab of the Ribbon. Triple-click to select any levels that are less than 24 pt and increase the font size accordingly.

3. To exit Slide Master mode, click the **Slide Master** tab and **Close Master**.

Windows

1. Click inside one of the text boxes in the Slide pane (see Figure A3.1A). Click the **Design | Themes | More** arrow to display all of the themes. Mouse over the themes. The Live Preview feature allows you to see how the change affects the slide before actually

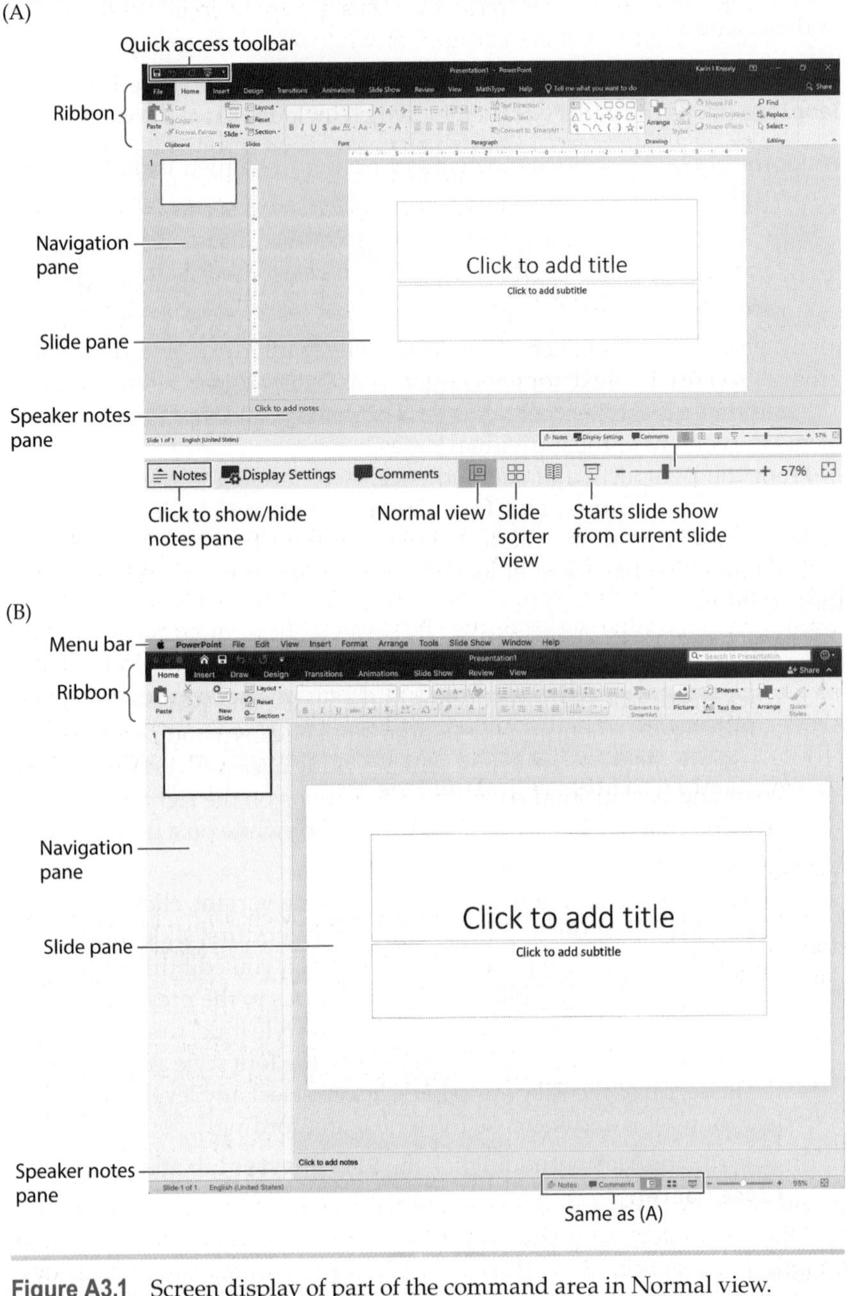
(A)
Quick access toolbar
Ribbon
Navigation pane
Slide pane
Speaker notes pane
Click to add title
Click to add subtitle
Click to add notes
Notes
Display Settings
Comments
57%
Click to show/hide notes pane
Normal view
Slide sorter view
Starts slide show from current slide
(B)
Menu bar
Ribbon
Navigation pane
Slide pane
Speaker notes pane
Click to add title
Click to add subtitle
Click to add notes
Same as (A)

Figure A3.1 Screen display of part of the command area in Normal view. (A) PowerPoint 2019 presentation, and (B) PowerPoint for Mac 2019 presentation. The status bar at the bottom of the screen has commands for changing views and showing and hiding the notes pane.

applying it. After you click a theme, mouse over the color palettes and background styles in the **Variants** group. Keep in mind that black on a light background or white on a dark background provides the best contrast.

2. To check the default font sizes for this theme variant, click **View | Master Views | Slide Master** and then scroll to the very first slide thumbnail in the navigation pane on the left. When you edit this slide, the changes will be applied to all of the slides in the presentation, not just to certain slide layouts. Click each style (title, first level text, second level text, and so on) to view the font style and size under **Home | Font**. Triple-click to select any levels that are less than 24 pt and increase the font size accordingly.

3. To exit Slide Master mode, click the **Slide Master** tab and **Close Master View**.

Choosing a slide size

Almost all TVs, computer screens, and projection screens have an aspect ratio of 16:9 (widescreen), which is why widescreen is the default slide size in PowerPoint 2019 and PowerPoint for Mac 2019. Widescreen, as the name implies, increases the width of the slide, so you have more room for content. However, if a slide formatted in widescreen is projected in the older 4:3 format, the sides of the slide will be cut off. For this reason, it's a good idea to find out beforehand how the slides at the presentation site will be displayed. If necessary, the slide size can be changed in PowerPoint by clicking **Design | Customize | Slide Size ▼**.

Views

The most frequently used views can be accessed by clicking a button on the status bar at the bottom of the screen (see Figure A3.1). Most of the time, you will be working in **Normal** view. **Normal** view consists of a Navigation pane, a Slide pane whose layout and content can be customized, and a Notes pane where you can write notes for your presentation, which will not be seen by the audience. The Notes pane can also be used to provide explanatory information about each slide if the slide deck will be shared. There is no limit to the amount of text that can be entered on the Notes pane.

   Slide Sorter view is useful for evaluating the overall appearance of a presentation because it displays thumbnails of all of the slides in the presentation, complete with slide transitions. You cannot edit the content of an individual slide in **Slide Sorter** view, but you can rearrange the slides. To select a slide to move, copy, or delete, single-click it. To edit the content of the slide, double-click it to return to **Normal** view. Slides can also be rearranged in the Navigation pane in **Normal** view.

Slide layouts

Presentations in biology usually follow the Introduction-Body-Closing format. In the Introduction, the speaker guides the audience from information that is generally familiar to them to new information, which is the focus (Body) of the talk. The takeaway message for the audience and any acknowledgments are given in the Closing.

Title slide   The default first slide in a new presentation is the Title Slide (see Figure A3.1). Follow the instructions in the placeholders to add text or add custom content. The same content will appear on the slide thumbnails on the left.

The Title Slide is going to set the tone for your presentation. A catchy title and interesting visuals are likely to capture your audience's attention. If possible, allude to the main benefit your audience will gain from listening to your presentation.

Adding slides   To insert a slide after the title slide, click below the slide thumbnail in the Navigation pane. A red horizontal line shows where the next slide will be inserted. Click **Home | Slides | New Slide ▼** and choose a slide layout.

The basic slide layout options allow you to arrange text and content (pictures, charts, tables, videos, and graphics) in one or two columns. The placeholder text boxes, however, can be moved, resized, and deleted as needed. Alternatively, if you decide that a different layout would work better after you've already added content to a slide, click **Home | Slides | Layout ▼** and select a different layout.

The last slide   A professional slide show has a definitive ending. End the show with an acknowledgments slide or add a slide that invites the audience to ask questions.

Reusing slides from other presentations

Slides from other presentations can be inserted in a new presentation. In Windows, the theme from the reused slide can be applied to other slides in the new presentation, or the reused slide can adopt the theme of the new presentation.

Mac

1. In the Navigation pane of the presentation you are working on, click where you want the reused slides to be inserted. This location will be marked by a horizontal red line.

2. To import *all of the slides* in the source presentation, click **Insert | New Slide ▼ | Reuse Slides** and navigate to a

PowerPoint presentation you want to reuse. Click **OK**. The imported slides will automatically adopt the theme of the target presentation.

3. To import *only certain slides* from another presentation, open that presentation, and click the slides to copy while holding down the ⌘ key. Click **Home | Copy** or press ⌘**+C**. In the target presentation, click **Home | Paste** or press ⌘**+V** to apply the target theme to the reused slides. Click **Home | Paste ▼ | Keep Source Formatting** if you intend to change the theme of the target presentation to that of the reused slides.

Windows

1. In the presentation you are working on, click **Insert | New Slide ▼ | Reuse Slides**. In the Reuse Slides pane on the right, navigate to a PowerPoint presentation you want to open. Thumbnails of the slides in the selected presentation will then be shown in the Reuse Slides pane.

2. In the Navigation pane on the left, click where you want the reused slide to be inserted. This location will be marked by a horizontal red line. In the Reuse Slides pane on the right, click the thumbnail of the slide to be reused.

3. The theme of the presentation you are working on will be applied to this slide unless you click the **Keep source formatting** checkbox at the bottom of the Reuse Slides pane. If you prefer to apply the theme of the reused slide to your current presentation, right-click the reused slide and select **Apply Theme to All Slides**.

Adding animation

Without animation, all of the objects on a slide are displayed for as long as the slide is displayed. **Animation** causes individual objects on a slide to move. Speakers use animation to help the audience focus on one element at a time, which is only revealed when the speaker is ready to talk about it. This technique is often used to list objectives or summarize conclusions.

If you decide to use animations, use the same animation throughout your presentation and use it selectively for emphasis. *You want your audience to be focused on what you are saying, not be distracted by the special effects!*

Mac   To animate an object, click it, and then select an animation from the **Animations** tab. Click the respective down arrows for the entrance and exit animations to display categories with different levels of excitement. For a scientific audience, it's probably best to stick to the basic animations, not *just* because they appear and disappear quickly.

To adjust the direction and sequence of the animation effect, click the **Effect Options** button. Other commands on the **Animations** tab allow you to customize how the animation will start (on click or connected to the previous animation), the duration, and delay time. To preview your animation, click the **Slide Show** button on the right side of the status bar (see Figure A3.1B). To delete an animation, click **Animations | Animation Pane**, select the animation to delete, and click the red X. To change the order in which the animations are played, use the up and down arrows on the Animation pane.

Windows   To animate an object, click it, and then select an animation from the **Animations** tab. Click the **More** arrow next to the animations and preview the options by clicking the respective buttons. Notice that there are both entrance and exit animations. More effects, with increasing levels of excitement, can be selected at the bottom of the options box. For a scientific audience, it's probably best to stick to the basic animations, not *just* because they appear and disappear quickly.

To adjust the direction and sequence of the animation effect, click the **Effect Options** button. Other commands on the **Animations** tab allow you to customize how the animations will start (on click or connected to the previous animation), the duration, delay time, and order in which they will be played. To delete an animation, click it and choose **Animations | Animation | None**. To preview your animation, click the **Slide Show** button on the right side of the status bar (see Figure A3.1A).

Importing Excel graphs

Excel charts can be imported into PowerPoint with full editing capability using the standard copy and paste commands. However, the drawback is that Excel's default 10 pt font makes the numbers and text close to unreadable for the audience. If you don't need to edit the chart in PowerPoint, then a better option is to paste the graph as a picture. To do so in the Mac OS, click **Home | Paste ▼ | Paste as Picture**. To do so in Windows, click **Home | Paste ▼ | Picture**. The font size automatically increases when you drag on one of the corners to enlarge the picture.

Working with shapes

Microsoft PowerPoint and Word offer a wide variety of shapes with which to make pointers, line drawings, and simple graphics. You can also make unique shapes by combining, coloring, and editing preexisting shapes. Begin by clicking **Insert | Illustrations | Shapes** and selecting a shape.

Lines   Lines (including arrows) are useful as pointers. After clicking a line from the shapes gallery and moving the mouse over the Slide pane to display crosshairs, hold down the left mouse button, drag to where you want the line to end, and release the mouse button. To make perfectly horizontal or vertical lines, hold down the **Shift** key while holding down the left mouse button and drag. In general, when it's important that a shape retain perfect proportions (e.g., circle not oval; square not rectangle), hold down the **Shift** key while drawing or resizing the shape with the mouse.

To change the angle or length of the line, click the line and then position the mouse pointer over one of the end points of the line to display a two-headed arrow. Hold down the left mouse button, drag to where you want the line to end, and release the mouse button.

To format the line, click it and select an attribute on the **Shape Format** tab in the Mac OS or **Drawing Tools Format** tab in Windows. Customize line color, weight, dash type, and arrow type, and, if you want, add special effects to give the line a three-dimensional look.

Text   To add text to your drawing, click **Insert | Text | Text Box**. Position the mouse pointer where you want the text box to appear, hold down the left mouse button, drag to enlarge the text box to the size you think you'll need (you can resize it later), and release the mouse button. Type text inside the box. In the Mac OS, change the font face and size by selecting the text and then changing the attributes on the **Home** tab. In Windows, right-click the text and use the buttons on the mini toolbar. Or add special effects using the buttons on the **Shape Format** tab in the Mac OS or **Drawing Tools Format** tab in Windows.

Text can also be added to two-dimensional shapes, such as circles, triangles, and squares. With the shape selected, just start typing. To format the text, use the options on the **Home** tab or on the **Shape Format** tab (Mac) or the **Drawing Tools Format** tab (Windows).

Group shapes   When several objects make up a graphic, it may make sense to group them as a unit. Grouping allows you to copy, move, format, and animate all the objects in the graphic at one time. To group individual objects into one unit, click each object while holding down the **Shift** key. Then,

- **Mac** Release the **Shift** key and click **Group Objects** on the **Shape Format** tab. One set of selection handles surrounds the entire unit when the individual objects are grouped. To ungroup, simply select the group and click **Group | Ungroup**. Grouping is temporary, whereas merging is permanent (see the next section).
- **Windows** Release the **Shift** key and click **Group** under **Drawing Tools Format | Arrange**. One set of selection handles surrounds the entire unit when the individual objects are grouped. To ungroup, simply select the group and click **Group | Ungroup**. Grouping is temporary, whereas merging is permanent (see the next section).

Merge shapes This command makes it possible to make unique shapes by permanently grouping standard shapes. For example, if you created a cartoon of a newly discovered organism, and you want to reuse the entire cartoon, not just the individual shapes, shift-click the individual objects. In the Mac OS, click **Shape Format | Merge Shapes**. In Windows, click **Drawing Tools Format | Insert Shapes | Merge Shapes**. *Note that once a presentation containing a merged shape has been saved, the shape cannot be unmerged.*

Align shapes When there are multiple objects on a slide, alignment guides appear when you drag one of the objects. **Alignment guides** are vertical and horizontal lines that show you the edges of other objects with which you may wish to align the selected object.

Alternatively, you can use the **Align** button to align selected objects by their top, bottom, right, or left edges or center them vertically or horizontally. To do so, shift-click each object that you want to line up. Release the **Shift** key, and then:

- **Mac** Click **Shape Format | Align ▼**, and select one of the alignment options.
- **Windows** Click **Drawing Tools Format | Arrange | Align ▼**, and select one of the alignment options.

Finally, to move objects just a fraction of a millimeter from their current position, click the object, hold down the ⌘ (**Cmd**) key (Mac) or the **Ctrl** key (PC), and use the arrow keys to nudge the object exactly where you want it on the slide.

Adding videos

To insert a video file on your slide:

- **Mac** Click **Insert | Video** inside a text box. Choose either **Movie Browser, Movie from File,** or **Online Movie**.
- **Windows** Click **Insert | Media | Video**, and choose either **Online Video** or **Video on My PC**.

When using online videos, you have the choice of inserting the actual video file or embedding the code on your slide. Embedding reduces the size of your PowerPoint presentation, but requires a reliable internet connection to play the video. To embed a YouTube video:

1. Go to www.youtube.com and find the desired video.
2. Below the video frame, click **Share** and then **Embed**. Copy the embed code.

3. Paste the embed code into the text box in PowerPoint. Click **Insert** (Mac) or ⇨ (Windows) to exit the dialog box. A box will be inserted on the slide, which you can size and reposition as needed.

4. To preview the video, click **Play** (Mac) or click the box and select **Preview | Play** from the **Video Tools Format** tab (Windows).

Text formatting shortcuts

Mac   To insert symbols (Greek letters, mathematical symbols, pictographs, etc.), click **Insert | Symbol**. Click a category (or type the name of the symbol into the search box) and double-click the symbol you want. Exit the dialog box. The symbol will be set in the same font as the text font in your PowerPoint presentation.

In PowerPoint, it is not possible to make shortcut keys for symbols, but it is possible to program them in AutoCorrect. To do so, copy the symbol and then click **Tools | AutoCorrect Options**. On the **AutoCorrect** tab, paste the symbol into the **With** box (it is not automatically entered, as it is in Word). Type a simple keystroke combination in the **Replace** box. Click the + sign at the bottom of the box, and then exit AutoCorrect.

A long expression such as 2-nitrophenyl β-D-galactopyranoside can also be replaced with a shorter one using AutoCorrect, but it is still not possible to italicize scientific names of organisms automatically as it is in MS Word's AutoCorrect.

Windows   To insert symbols (Greek letters, mathematical symbols, Wingdings, etc.), click **Insert | Symbols | Symbol**, and click the desired symbol and close the box. For a uniform look, select the symbol on your PowerPoint slide and apply the text font that is used in your PowerPoint presentation.

In PowerPoint, it is not possible to make shortcut keys for symbols, but it is possible to program them in AutoCorrect. To do so, copy the symbol and then click **File | Options | Proofing | AutoCorrect Options**. On the **AutoCorrect** tab, paste the symbol into the **With** box (it is not automatically entered, as it is in Word). Type a simple keystroke combination in the **Replace** box and then exit AutoCorrect.

Interestingly, many symbols have already been entered on the **MathAutoCorrect** tab, but these keystrokes only work inside equation boxes. However, you can copy the symbols and their corresponding keyboard shortcuts from the **MathAutoCorrect** tab and paste them into their respective boxes on the **AutoCorrect** tab.

A long expression such as 2-nitrophenyl β-D-galactopyranoside can also be replaced with a shorter one using AutoCorrect, but it is still not possible to italicize scientific names of organisms automatically as it is in MS Word's AutoCorrect.

A3.4 Delivering Presentations

A PowerPoint presentation can be "delivered" in several ways:

- Attaching the file to an email
- Posting the file to a website
- Recording the slide show and converting it to a video (mp4 file)

However, most presentations are delivered in person. Table A3.1 gives some presentation tips. See Chapter 10 for detailed instructions on planning and delivering oral presentations.

Presenter View

Presenter View is seen only by the presenter (Figure A3.2). It helps the speaker stay on track by displaying speaking prompts and a preview of the next slide without making the slide that the audience sees text-heavy. Audience members see only the current slide with the content that you specifically chose to help them focus on the main point.

Rehearsal

Mac To use Presenter View on your computer when you are rehearsing your presentation, press **Option+Return** on the keyboard or click **Slide Show | Presenter View** on the Ribbon. As shown in Figure A3.2A, there is a timer on the left and a clock on the right above the current slide. The timer starts as soon as you start Presenter View, but it can be paused or reset as needed. Notes can be added in this view, but if you need to edit a slide, press **Esc** to exit. Changes to slides can only be made in **Normal** view.

TABLE A3.1 Presentation tips	
Do	**Don't**
Keep the wording simple	Write every word you're going to say on the slide
Make the text large and legible	Use backgrounds that make text and images hard to read
Strive for a consistent look (use the same font and format for each slide)	Use distracting animations and slide transitions, or sound effects
Include visuals that complement and support the auditory information	Include tables when you can show the trend better with a graph
Allow enough time for each slide	Talk endlessly without referring to a visual
	Rush through the slides without mentioning the takeaway message about each one

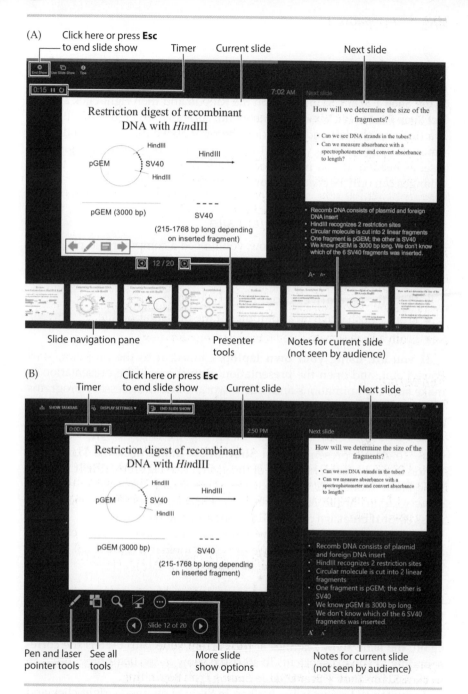

Figure A3.2 Presenter View is seen only on the presenter's computer screen. This view shows, on the left, the current slide that is seen by the audience; speaker notes for that slide bottom right; and the next slide top right. Presenter tools allow the speaker to write on the slides, turn the mouse into a laser pointer, navigate among slides, zoom in, and black out the screen. (A) PowerPoint for Mac 2019 screen shot. (B) PowerPoint 2019 screen shot.

Windows   To use Presenter View on your computer when you are rehearsing your presentation, click **Alt+F5** on the keyboard or click **Slide Show | Start Slide Show** on the Ribbon and then right-click and select **Show Presenter View** from the menu. As shown in Figure A3.2B, there is a timer on the left and a clock on the right above the current slide. The timer starts as soon as you start Presenter View, but it can be paused or reset as needed. If you need to edit a slide or the notes, press **Esc** to exit. Changes can only be made in **Normal** view.

Resources at the presentation site

To deliver a PowerPoint slide show, you'll need a computer and projection equipment as well as a screen on which to display the slides. Assuming these requirements are met, there are several ways to run the presentation:

- From the hard drive of your laptop computer
- From a flash drive on a computer at the presentation site
- From a Web server

If you are using your own laptop, connect it to the projector, start PowerPoint, and open the presentation. Run through the presentation to make sure the animations and videos work. Close unnecessary programs that might cause the computer to run slower or the audience to become distracted.

Mac   To start a presentation in **Slide Show mode**, click the **Slide Show** button in the lower right corner of the screen (see Figure A3.1B). If you exit Slide Show mode and then click the **Slide Show** button again, the slide show resumes at the current slide. To start the show at the beginning, press ⌘**+Shift+Return** or click **Slide Show | Play from Start**.

To switch to Presenter View when in Slide Show mode, click the **Slide Show options** button on the presenter toolbar (Figure A3.3), and select **Use Presenter View** from the menu.

Windows   To start a presentation in **Slide Show mode**, click the **Slide Show** button in the lower right corner of the screen (see Figure A3.1A). If you exit Slide Show mode and then click the **Slide Show** button again, the slide show resumes at the current slide. **Shift+F5** is the corresponding keyboard shortcut. To start the show at the beginning, press **F5** or click **Slide Show | Start Slide Show | From Beginning**.

To switch to Presenter View when in Slide Show mode, right-click and select **Show Presenter View** from the menu.

If the presentation site has a computer, but it is not hooked up to the internet, save your presentation file and any linked videos on a flash drive and carry it with you. Assuming that PowerPoint is installed on

Slide show options

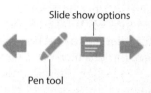

Pen tool

Figure A3.3 Presenter toolbar in the Mac OS appears at the bottom of the current slide in Slide Show mode and Presenter View. The Slide Show options button opens a menu that lets you navigate to another slide, make the screen black or white to focus audience attention away from the projection screen, end the slide show, and toggle between Slide Show mode and Presenter View. The Pen tool allows you to write on the slides. The writing is temporary and will not be saved after you navigate to another slide.

the presentation room's computer, plug the flash drive into the USB port, download the files to desktop, and start your presentation as described previously.

To run a presentation from a Web server, first find out if you have access to the internet at the presentation site. If so, save your file to a remote server, such as Google Drive or Dropbox. Similarly, if your college or university has networked computers, you can prepare your presentation on a networked computer in your room, save the file in your Netspace, and then access the file from another networked computer in the room where you will hold your talk.

Presenter tools

When you give your presentation, some of the actions you will perform are
- Starting and ending your slide show
- Moving forward through the slides
- Navigating to selected slides
- Pointing to specific slide elements
- Writing on the slides for emphasis

PowerPoint offers many presenter tools that can be accessed by keyboard and/or mouse on the computer that is connected to the projector. To use these tools, however, you have to be standing near the computer.

If you prefer to move around during your talk, a wireless presenter allows you to change slides from anywhere in the room. This device usually consists of a USB receiver that you plug into the computer's USB port, and the remote control, which has buttons for moving forward and backward, pausing, or stopping the slide show. The remote control may also have a built-in laser pointer.

Mac   To advance to the next slide:

- On the keyboard, press **Return** or the space bar.
- Using the mouse, click the forward arrow in the toolbar or click the left mouse button. In Presenter View, you can also click the current slide.

If a slide has animations, these procedures will advance to the next animation. To go back to the previous slide or animation:

- Press **Delete**.
- Click the back arrow in the toolbar.

To navigate to a specific slide in Slide Show mode, click **Slide Show options | By Title**, and select the slide from the menu. In Presenter View, click a slide from the Slide Navigation pane at the bottom of the screen (see Figure A3.2A).

To turn the mouse into a laser pointer or to write on the slides, click the **Pen tool**.

To end a slide show, press **Esc** or click the **End Show** button at the top of the Presenter View window.

Windows   To advance to the next slide:

- On the keyboard, press **Enter** or the space bar.
- Using the mouse, click the forward arrow in the toolbar, click the left mouse button, or click either the current or the next slide.

If a slide has animations, these procedures will advance to the next animation. To go back to the previous slide or animation:

- Press **Backspace**.
- Click the back arrow in the toolbar.
- Click the right mouse button to display a menu of navigation options.

To navigate to a specific slide in Slide Show mode, right-click and select **See all slides** from the menu. In Presenter View, click the **See all slides** button in the presenter toolbar at the bottom of the screen (see Figure A3.2B) and then click the desired slide.

To turn the mouse into a laser pointer or to write on the slides, click **Pen and laser pointer tools**.

To end a slide show, press **Esc** or click the **End Slide Show** button at the top of the Presenter View window.

Speaker notes, handouts, and sharing electronic presentations

The Notes pane is useful for listing the talking points for each slide and for providing explanatory information in handouts and for other presenters who may be reusing your slide show. If you are comfortable referring to your notes on a computer screen in Presenter View, then there is no need to print out speaker notes. However, if you feel more comfortable presenting with notes in hand, follow the instructions below to print a large thumbnail of the slide along with notes. When printing on a black-and-white printer, select **Pure Black and White** to avoid printing out the background of the slides. The same instructions can be used to produce printed handouts or PDFs if posting online.

Mac   Click **File | Print | Show Details | Layout: Notes**.

Windows   Click **File | Print | Print Layout | Notes Pages** or **File | Export | Create Handouts**.

Because high resolution images can increase the size of a PowerPoint presentation significantly, you may wish to compress the pictures first as follows:

Mac   Click **Picture Format | Compress Pictures** and then choose an option from the menu.

Windows   Click **Picture Tools Format | Adjust | Compress Pictures** and then choose an option from the menu.

Other ways to share slide decks are described in Table 10.2.

Recording a presentation

PowerPoint presentations can be narrated, saved as a video file, and posted for remote viewing. Start by writing the talking points for each slide in the Notes Pane. These talking points will be displayed at the top of the screen in recording mode, just below the built-in webcam (or an external webcam centered above the monitor). If you are an experienced speaker, a bulleted list of talking points is probably sufficient. If you tend to get nervous speaking in public, it may be better to write full sentences in the Notes Pane. That way, you don't have to worry about your mind going blank during the recording.

When you are ready to record your presentation, click **Slide Show | Set Up | Record Slide Show**. Figure A3.4 shows some of the features you may use in recording mode. First, check the sound quality of the microphone. Built-in microphones may be good enough for communicating informally, but external microphones are preferred for higher audio quality. External mics can be positioned closer to your mouth, which improves voice quality and reduces background noise. Second, decide if you would like the audience to see you while you are speaking or just hear you as they look at the slides. Turn off the webcam if you only want to record audio. If you choose to turn on the webcam for one or more slides, be

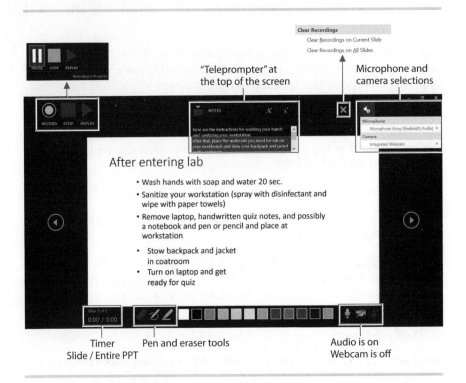

Timer — Slide / Entire PPT — Pen and eraser tools — Audio is on, Webcam is off — "Teleprompter" at the top of the screen — Microphone and camera selections — Clear Recordings

Figure A3.4 Recording View in Windows is accessed by clicking **Slide Show | Set Up | Record Slide Show**. To save the narrated PowerPoint as an mp4 video, click **File | Export | Create a Video | Use recorded timings and narrations**.

aware that your picture may cover text or images on the slide. (As of September 2020, the webcam option is available only in Windows.)

Click **Record** and then begin speaking. If you make a mistake, simply click **Stop** and record again. Clicking the X gives you the option of clearing the recording on the current slide or on all slides. When advancing to the next slide, wait until the slide is displayed before continuing your narration. The audio will automatically be synced to each slide in the final video file.

The pen tools can be used to write on the slides during the presentation. The markup will be saved when you save the file. To save the recorded file as an mp4 video, click **File | Export | Create a Video | Use recorded timings and narrations**.

Go to **macmillanlearning.com/knisely6e**
and select **"Student Site"** to access
samples, template files, and tutorial videos

Bibliography

Academic skills

Bucknell University. The Teaching & Learning Center. c2020. Student Learning Support http://www.bucknell.edu/LearningCenter

Clarion University Writing Center. c2020. Paraphrasing 101 [accessed 2020 Oct 2] https://www.clarion.edu/academics/student-success-center/writing-center/paraphrasing.pdf

Indiana University Bloomington Writing Tutorial Services. c2020. Plagiarism: What it is and how to recognize and avoid it. [accessed 2020 Oct 1] https://wts.indiana.edu/writing-guides/plagiarism.html

Novak JD, Cañas AJ. 2008. The theory underlying concept maps and how to construct and use them, Technical Report IHMC CmapTools 2006-01 Rev 01-2008, Florida Institute for Human and Machine Cognition; [updated 2008 Jan 22; accessed 2020 Oct 2]. http://cmap.ihmc.us/docs/pdf/TheoryUnderlyingConceptMaps.pdf

Palmer-Stone D. 2010. Mind-mapping. Victoria (BC), Canada: University of Victoria, Counselling Services; [accessed 2020 Sept 30]. https://www.uvic.ca/services/counselling/assets/docs/MindMapping.pdf

Purdue OWL. c1995-2020. Paraphrase: Write it in your own words. The Writing Lab & The OWL at Purdue and Purdue University. [2020 Sept 30] https://owl.purdue.edu/owl/research_and_citation/using_research/quoting_paraphrasing_and_summarizing/paraphrasing.html

Shaffer S. 2016. Active reading. iStudy for Success! Penn State University, IT Learning and Development. [updated 2017 Aug 8; accessed 2020 Oct 2]. http://tutorials.istudy.psu.edu/activereading/

Stanford University Student Learning Programs. c2020. Time management, note-taking, and reading tips and tools. [accessed 2020 Oct 2] https://studentlearning.stanford.edu/academic-skills/tips-and-tools

University of Victoria. 2020. Reading and concept mapping learning module. Victoria (BC), Canada: University of Victoria, Counselling Services; [accessed 2020 Sept 30]. https://www.uvic.ca/services/counselling/assets/docs/Learning%20Module%20Reading%20and%20Concept%20Mapping.pdf

University of Victoria. 2020. Survey reading learning module. Victoria (BC), Canada: University of Victoria, Counselling Services; [accessed 2020 Sept 30]. https://www.uvic.ca/services/counselling/assets/docs/learning%20module%20Survey%20Reading.pdf

Finding information and evaluating credibility

American Library Association. 1989. Presidential Committee on Information Literacy: Final Report. [accessed 2020 Aug 6]. http://www.ala.org/acrl/publications/whitepapers/presidential

Barton CJ, Merolli MA. 2019. It is time to replace publish or perish with get visible or vanish: opportunities where digital and social media can reshape knowledge translation. *British Journal of Sports Medicine* 53(10): 594–598

Benedictine University. Fake news: Develop your fact-checking skills: Fact-checking sites and plug-ins [updated 2020 Jun 30; accessed 2020 Jul 5] https://researchguides.ben.edu/fake-news

Colmers IN, Paterson QS, Lin M, Thoma B, Chan TM. 2015. The quality checklists for health professions blogs and podcasts. *The Winnower* 6:e144720.08769. doi:10.15200/winn.144720.08769

Goldman SR, Braasch JLG, Wiley J, Graesser AC, Brodowinska K. 2012. Comprehending and learning from internet sources: Processing patterns of better and poorer learners. *Reading Research Quarterly* 47(4): 356–381

How Google Search works. 2020. [accessed 2020 Jul 10] https://www.google.com/search/howsearchworks/

James S. 2020. Tips on using Science Twitter during COVID-19. [2020 April 15; accessed 2020 Sept 4] https://scicomm.plos.org/2020/04/15/tips-on-using-science-twitter-during-covid-19/

JSTOR's Text Analyzer. *This tool can be used to find related articles. Navigate to the JSTOR database on your library's webpage (this database is by subscription). Click* **Tools | Text Analyzer**. *Select a document or an image from your computer for analysis. The tool returns results based on the main concepts identified in the document. The left panel of the results page displays other terms identified in the document. The weight of each prioritized term can be changed with a slider, and terms can be deleted or added to refine the search. Articles in JSTOR can be read online, download as a PDF, shared, or saved to JSTOR's Workspace.*

Kiili C, Leu DJ. 2019. Exploring the collaborative synthesis of information during online reading. *Computers in Human Behavior* 95: 146–157

Kohnen AM, Mertens GE. 2019. "I'm always kind of double-checking": Exploring the information-seeking identities of expert generalists" *Reading Research Quarterly* 54(3): 279–297. doi:10.1002/rrq.245

Madison College Libraries. Information literacy: Guide for students: What is information literacy? [updated 2020 March 25; accessed 2020 Jul 5]. https://libguides.madisoncollege.edu/InfoLitStudents

Mount Saint Vincent University. LIBR 2100 course materials. [updated 2020 Jun 24; accessed 2020 Jul 20]. https://libguides.msvu.ca/oldLIBR2100 *This website has many useful tutorials on the research process, article databases, Google Scholar, strategies to enhance reading comprehension, plagiarism, and other topics.*

PubMed Help. c2005–. Bethesda (MD): National Center for Biotechnology Information (US) [updated 2020 Mar 31; accessed 2020 Jul 18]. https://www.ncbi.nlm.nih.gov/books/NBK3827/

UC Libraries: University of California. A cycle of revolving research [accessed 2020 Jul 14] https://library.ucmerced.edu/rpt/knowledge_cycle_8

Microsoft Office

LinkedIn Learning. PowerPoint tips and tricks. [accessed 2020 Aug 23] https://www.linkedin.com/learning/powerpoint-tips-and-tricks/display-the-recording-tab?u=50365505

Microsoft Support. c2020. Office help and training. [accessed 2020 Dec 7]. https://support.office.com/en-us

Miscellaneous

Martin B. 2011. Doing good things better. Ed (Sweden): Irene Publishing. [accessed 2020 Oct 2]. Available free online at: http://www.bmartin.cc/pubs/11gt/

Nijhuis M. 2011. Crisis in the caves. *Smithsonian* 42(4): 66–74

Reeder DM, et al. 2012. Frequent arousal from hibernation linked to severity of infection and mortality in bats with white-nose syndrome. *PLOS* 7(6): e38920

Oral presentations

Alley M. c2016. Assertion-evidence approach. State College (PA): The Pennsylvania State University; [accessed 2020 Oct 11]. http://www.craftofscientificpresentations.com/assertion-evidence-approach.html

D'Arcy J. 1998. Technically speaking: A guide for communicating complex information. Columbus: Battelle Press. 270 p.

Fegert F, Hergenröder F, Mechelke G, Rosum K. 2002. Projektarbeit: Theorie und Praxis. Stuttgart: Landesinstitut für Erziehung und Unterricht Stuttgart. 226 p.

Hailman JP, Strier KB. 2006. Planning, proposing, and presenting science effectively: A guide for graduate students and researchers in the behavioral sciences and biology, 2nd ed. Cambridge: Cambridge University Press. 248 p.

McConnell S. 2011. Designing effective scientific presentations. [uploaded 13 Jan 2011; accessed 2020 Oct 11]. https://www.youtube.com/watch?v=Hp7Id3Yb9XQ

Nathans-Kelly T, Nicometo CG. 2014. Slide rules: Design, build, and archive presentations in the engineering and technical fields [e-book]. Hoboken (NJ): Wiley-IEEE Press. 219 p.

Popular science magazines

Atlantic Monthly, Discover, EurekAlert!, National Geographic, New York Times, New York Times Magazine, New Scientist, Popular Science, Science (AAAS), Scientific American, Smithsonian, Technology Review, Wired, among others

Posters

Alley M. c1997-2016. Scientific Posters. State College (PA): The Pennsylvania State University; [accessed 2020 Sept 9]. http://www.craftofscientificposters.com/

Everson K. The scientist's guide to poster design. [accessed 2020 Sept 9]. http://www.kmeverson.org/academic-poster-design.html

Hess GR, Tosney K, Liegel L. c2014. Creating effective poster presentations. [accessed 2020 Sept 9]. https://go.ncsu.edu/posters

Poster creation and presentation. c2020. State College (PA): The Pennsylvania State University; [last updated 2019 Dec 3; accessed 2020 Sept 9]. https://guides.libraries.psu.edu/posters

Purrington CB. c1997–2016. Designing conference posters. [accessed 2020 Sept 9]. http://colinpurrington.com/tips/poster-design

Woolsey JD. 1989. Combating poster fatigue: How to use visual grammar and analysis to effect better visual communications. *Trends in Neuroscience* 12(9): 325–332.

Reading and writing about biology

Gillen CM. 2007. Reading Primary Literature. San Francisco: Pearson/Benjamin Cummings. 44 p.

Hofmann AH. 2019. Scientific Writing and Communication: Papers, Proposals, and Presentations, 4th ed. New York: Oxford University Press. 768 p.

Lannon JM, Gurak LJ. 2020. Technical Communication, 15th ed. New York: Pearson.

McMillan VE. 2021. Writing Papers in the Biological Sciences, 7th ed. New York: Macmillan Learning.

Pechenik JA. 2016. A Short Guide to Writing about Biology, 9th ed. New York: Pearson Education. 288 p.

VanAlstyne JS. 2005. Professional and technical writing strategies, 6th ed. New York, NY: Pearson Longman. 752 p.

Reading journal articles

Lockman T. 2012. How to read a scholarly journal article. Kishwaukee College Library. [accessed 2020 Oct 2] https://www.youtube.com/watch?v=EEVftUdfKtQ

University of Minnesota Libraries. 2014. How to read and comprehend scientific research articles. [accessed 2020 Oct 2]. https://www.youtube.com/watch?v=t2K6mJkSWoA

Revision

Every B. c2006–2015. My 15 best proofreading tips. BioMedical Editor [accessed 2020 Sept 11]. http://www.biomedicaleditor.com/proofreading-tips.html

Koerber D. 2019. 100 Editing and proofreading tips for writers. [accessed 2020 Sept 11]. https://experteditor.com.au/blog/100-editing-and-proofreading-tips-for-writers/

The City University of New York School of Professional Studies Writing Fellows. 2020. CUNY Academic Commons. [accessed 2020 Sept 11]. https://bacwritingfellows.commons.gc.cuny.edu/proofreading/

The Writing Center at UNC Chapel Hill. 2020. Revising srafts. Chapel Hill (NC): University of North Carolina; [accessed 2020 Sept 11]. http://writingcenter.unc.edu/handouts/revising-drafts/

The Writing Center at UNC Chapel Hill. 2020. Editing and proofreading. Chapel Hill (NC): University of North Carolina; [accessed 2020 Sept 11]. http://writingcenter.unc.edu/handouts/editing-and-proofreading/

Science communication

American Association for the Advancement of Science. 2020. AAAS communication toolkit. [accessed 2020 Sept 3] https://www.aaas.org/resources/communication-toolkit

American Association for the Advancement of Science. 2020. Communication fundamentals. [accessed 2020 Oct 9] https://www.aaas.org/resources/communication-toolkit/communication-fundamentals

American Academy of Arts & Sciences. 2020. The public face of science in America: Priorities for the future. [accessed 2020 Aug 30] https://www.amacad.org/project/public-face-science

American Geophysical Union. 2020. SciComm toolkits. [accessed 2020 Aug 30] https://connect.agu.org/sharingscience/resources/tookits

Blum D, Knudson M, Marantz-Henig R, editors. 2005. A field guide for science writers: The official guide of the National Association of Science Writers. 2nd ed. [e-book] New York (NY): Oxford University Press USA

Bullock OM, Amill DC, Shulman HC, Dixon GN. 2019. Jargon as a barrier to effective science communication: Evidence from metacognition. *Public Understanding of Science* 28(7): 845–853. doi: 10.1177/0963662519865687

Fahnestock J. 1986. Accommodating science: The rhetorical life of scientific facts. *Written Communication* 3:275-296

Rakedzon T, Baram-Tsabari A. 2017. To make a long story short: A rubric for assessing graduate students' academic and popular science writing skills. *Assessing Writing* 32: 28–42

Yeo SK. 2015. Public engagement with and communication of science in a Web-2.0 media environment. White paper prepared for the American Association for the Advancement of Science (AAAS). [accessed 2020 Dec 8]. https://www. aaas.org/sites/default/files/content_files/public%20engagement%20social%20 media_Yeo_single.pdf

Scientific style and format
Council of Science Editors, Style Manual Committee. 2014. Scientific style and format: The CSE manual for authors, editors, and publishers. 8th ed. Reston (VA): The Council of Science Editors. 722 p.

Patrias K. 2007–. Citing medicine: the NLM style guide for authors, editors, and publishers. 2nd ed. Wendling DL, technical editor. Bethesda (MD): National Library of Medicine (US); [updated 2015 Oct 2; accessed 2020 Dec 7]. http:// www.nlm.nih.gov/citingmedicine

Peterson SM. 1999. Council of Biology Editors guidelines No. 2: Editing science graphs. Reston (VA): Council of Biology Editors. 34 p.

Peterson SM, Eastwood S. 1999. Council of Biology Editors guidelines No. 1: Posters and poster sessions. Reston (VA): Council of Biology Editors. 15 p.

Social media
Australian Critical Care Editorial 2019. Social media for researchers – beyond cat videos, over sharing, and narcissism. *Australian Critical Care* 32: 351–352

Diaz-Faes AA, Bowman TD, Costas R. 2019. Towards a second generation of 'social media metrics': Characterizing Twitter communities of attention around science. *PLOS ONE* 14(5): e0216408. https://doi.org/10.1371/ journalpone.0216408

Gierth L, Bromme R. 2020. Attacking science on social media: How user comments affect perceived trustworthiness and credibility. *Public Understanding of Science* 29(2): 230–247.

Hepatology Editorial 2018. Social media: Why AASLD and its members must lead the conversation. *Hepatology* 68:4–6

Lima DL, de Medeiros Lopes MAAA, Brito AM. 2020. Social media: friend or foe in the COVID-19 pandemic? Clinics 2020: 75:e1953. doi: 10.6061/clinics/2020/e1953

Mohammadi E, Thelwall M, Kwasny M, Holmes KL. 2018. Academic information on Twitter: A user survey. *PLOS ONE* 13(5): e0197265. https://doi.org/10.1371/ journalpone.0197265

Musculoskeletal Editorial 2017. Musculoskeletal science & practice and social media. *Musculoskeletal Science and Practice* 29: v-vi

Nature Cell Biology Editorial. 2018. Social media for scientists. *Nature Cell Biology* 20: 1329

Pew Research Center. July 2020. Americans who mainly get their news on social media are less engaged, less knowledgeable. [accessed 2020 Aug 7] https:// www.pewresearch.org/topics/social-media/

Regenberg A. 2019. Science and social media. *Stem Cells Translational Medicine* 00:1–4

Schuette S, Folk RA, Cantley JT, Martine CT. 2018. The hidden Heuchera: How science Twitter uncovered a globally imperiled species in Pennsylvania, USA. *PhytoKeys* 96: 87–97. https://doi.org/10.3897/phytokeys.96.23667

Twitter, Inc. 2020. How to use Twitter Lists. [accessed 2020 Sept 4] https://help. twitter.com/en/using-twitter/twitter-lists

Yeo SK, Liang X, Brossard D, Rose KM, Korzekwa K, Scheufele DA, Xenos MA. 2017. The case of# arseniclife: Blogs and Twitter in informal peer review. *Public Understanding of Science* 26(8): 937–952

Statistics

Barfield ET, Moser VA, Hand A, Grisel J. 2013. β-endorphin modulates the effect of stress on novelty-suppressed feeding. *Frontiers in Behavioral Neuroscience* 7: 1–7. doi: 10.3389/fnbeh.2013.00019

Barnett B, Kryczek I, Cheng P, Zou W, Curiel TJ. 2005. Regulatory T cells in ovarian cancer: biology and therapeutic potential. *American Journal of Reproductive Immunology* 54(6): 369–377. doi: 10.1111/j.1600-0897.2005.00330.x

Dunn OJ. 1961. Multiple comparisons among means. *Journal of the American Statistical Association* 56(293): 52–64.

Feenstra R, Hoffman B. 2015. How ice cream kills: Understanding cause and effect. [accessed 2020 Sept 2] https://www.decisionskills.com/blog/how-ice-cream-kills-understanding-cause-and-effect

Genome-Wide Association Study Fact Sheet. 2020. [accessed 2020 Nov 20]. https://www.genome.gov/about-genomics/fact-sheets/Genome-Wide-Association-Studies-Fact-Sheet

Ihaka R. 2017. The R project: A brief history and thoughts about the future. The University of Auckland. (interaktyvus) žiūrėta 4: 22.

Judge PG, Evans DW, Schroepfer KK, Gross AC. 2011. Perseveration on a reversal-learning task correlates with rates of self-directed behavior in nonhuman primates. *Behavioural Brain Research* 222(1): 57–65. https://doi.org/10.1016/j.bbr.2011.03.016

Korkmaz G, Kelling C, Robbins C, Keller SA. 2018. Modeling the impact of R packages using dependency and contributor networks. In 2018 IEEE/ACM International Conference on Advances in Social Networks Analysis and Mining (ASONAM) (pp. 511–514). IEEE.

Liang S, Carlin BP, Gelfand AE. 2009. Analysis of Minnesota colon and rectum cancer point patterns with spatial and nonspatial covariate information. *The Annals of Applied Statistics* 3(3): 943–962.

Pearl J. 2009. Causal inference in statistics: An overview. *Statistics Surveys* 3: 96–146. doi: 10.1214/09-SS057

R Core Team 2017. R: A language and environment for statistical computing. Vienna (Austria): R Foundation for Statistical Computing. https://www.r-project.org/

Teetor P. 2011. R cookbook: Proven recipes for data analysis, statistics, and graphics. O'Reilly Media, Inc. 438 p.

Wald HS, Myers KP. 2015. Enhanced flavor–nutrient conditioning in obese rats on a high-fat, high-carbohydrate choice diet. *Physiology & Behavior* 151: 102–110. doi: 10.1016/j.physbeh.2015.07.002

Wasserstein RL, Lazar NA. 2016. The ASA statement on p-values: context, process, and purpose. *The American Statistician* 70(2): 129–133. doi: 10.1080/00031305.2016.1154108

Wickham H, et al. 2019. Welcome to the Tidyverse. *Journal of Open Source Software* 4(43): 1686. https://doi.org/10.21105/joss.01686

Writing guides

Bullock R, Brody M, Weinberg F. 2021. The little seagull handbook, 4th ed. New York: W.W. Norton & Co., Inc. *This manual has a section on Council of Science Editors style.*

Hacker D, Sommers N. 2021. A pocket style manual, 9th ed. Boston: Bedford/St. Martin's. *This manual has a section on Council of Science Editors style and a section on evaluating sources. This writing guide is my favorite.*

Lunsford AA. 2020. The everyday writer, 7th ed. Boston: Bedford/St. Martin's. 608 p.

Lunsford AA. 2019. Easy writer, 7th ed. Boston: Bedford/St. Martin's. 416 p.

Lunsford AA. 2021. The new St. Martin's handbook, 9th ed. Boston: Bedford/St. Martin's.

Index